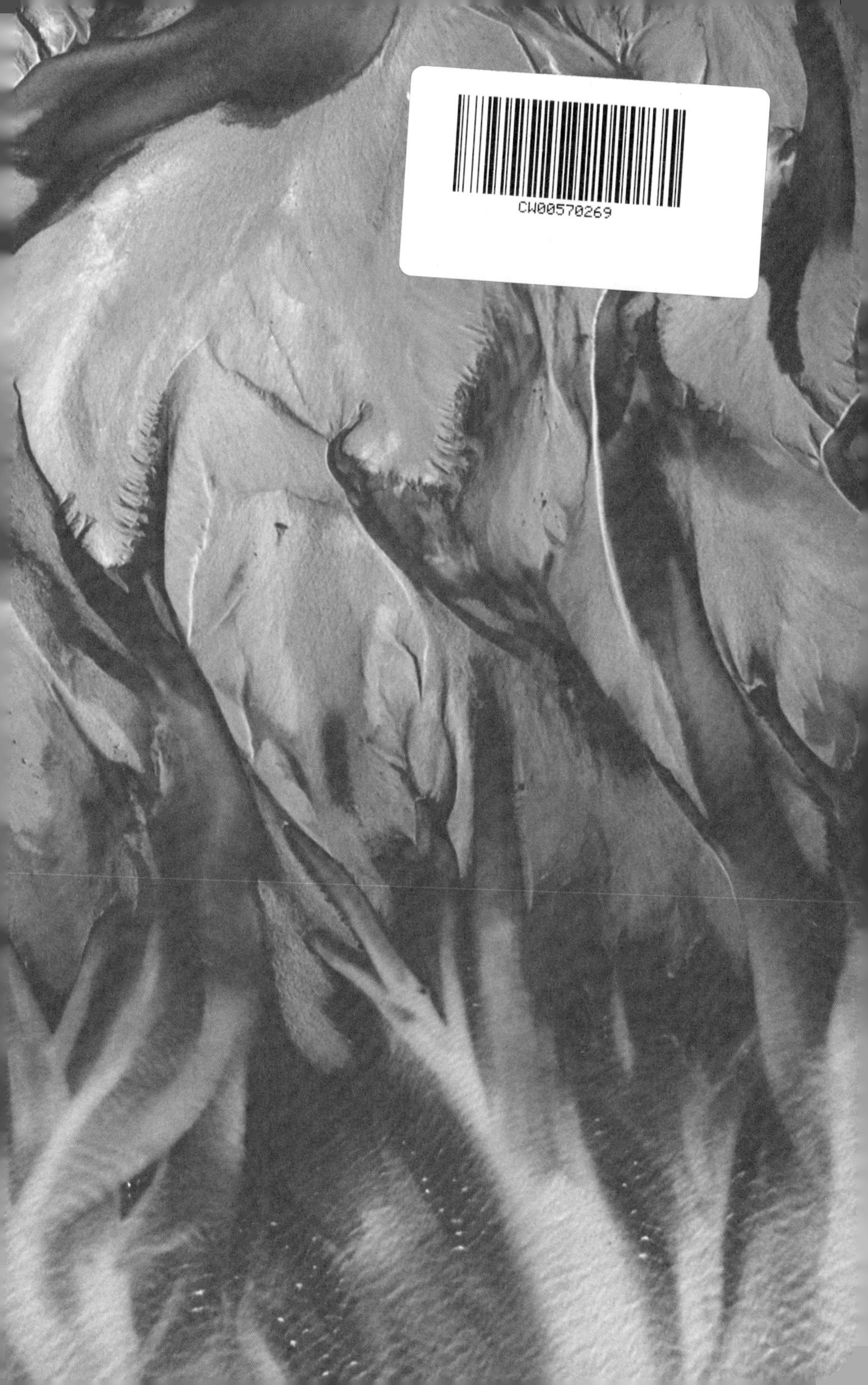
CW00570269

THIS JOURNAL BELONGS TO:

GRATITUDE
REVEALED

EACH AND EVERY ONE OF US CAN FIND SIMPLE ENTRY POINTS FOR PRACTICING GRATITUDE EACH AND EVERY DAY.

GRATITUDE REVEALED IS A JOURNEY YOU'LL TAKE INTO THE SCIENCE, MYSTERY, AND PURSUIT OF THE BUILDING BLOCKS OF BEING GRATEFUL. INCREASING GRATITUDE IS A PROVEN PATHWAY BACK FROM THE DISCONNECTION WE FEEL IN OUR LIVES; DISCONNECTION FROM OURSELVES, OUR PLANET, AND EACH OTHER.

WITH THESE FIFTEEN PRINCIPLES, YOU'LL EXPLORE WHAT GRATITUDE IS, WHY IT'S IMPORTANT, AND WHAT YOU CAN DO TO LIVE A MORE GRACIOUS LIFE. FROM WONDER AND CURIOSITY TO COURAGE AND GENEROSITY, YOU'LL EXAMINE DIFFERENT COMPONENTS OF EMOTIONAL WELLNESS AS YOU CONNECT YOUR COMMUNITY WITH TOOLS AND RESOURCES TO HELP THEM ALONG THE WAY.

WELCOME TO THE JOURNEY!

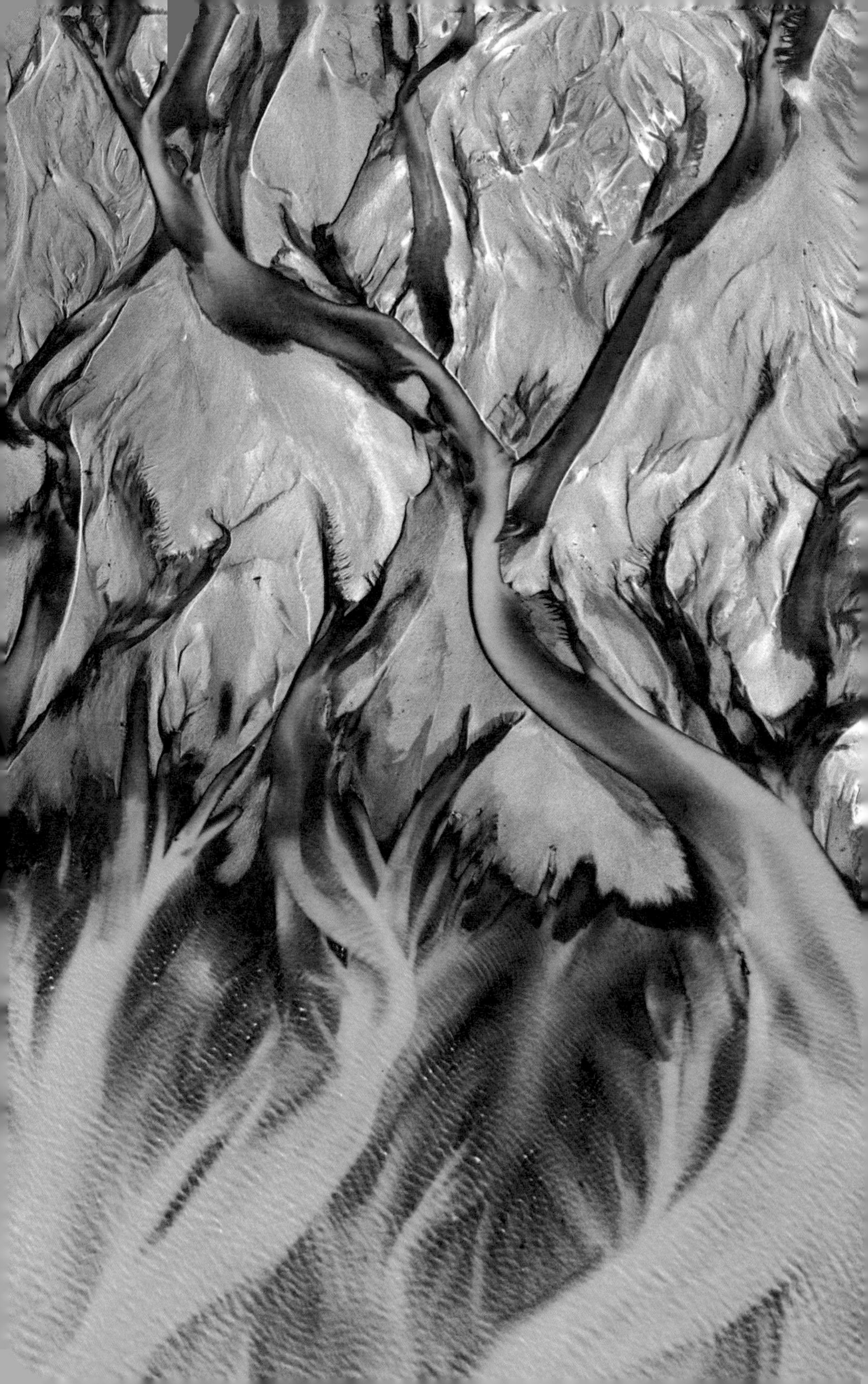

CONTENTS

CREATIVITY

CREATIVE PERSONALITIES LEAD TO POSITIVE TRAITS: People who have more creative personalities tend to be more independent, unconventional, open-minded, flexible, and spontaneous, as well as more self-confident and self-accepting.

CREATIVITY IS GOOD FOR STUDENTS: For example, in one study, at-risk youth who participated in arts programs ended up with better work habits, attitudes, and behavior.

CREATIVE PEOPLE ARE HEALTHIER: Studies suggest that participating in creative activities improves physical and psychological health and is even associated with stronger immune systems in HIV-positive patients.

CREATIVE PEOPLE ARE HAPPIER: Creativity is associated with greater well-being and life satisfaction and can help people maintain healthy relationships.

CREATIVITY ISN'T ALL-OR-NOTHING, AND IT'S NOT ONLY FOR ARTISTIC GENIUSES: What researchers refer to as "everyday creativity"—coming up with unique ideas that are useful or meaningful—is something that everyone is capable of and can improve at.

Research provided by UC Berkeley's Greater Good Science Center.

MEANINGFUL PHOTOS

TIME REQUIRED: 15 MINUTES DAILY FOR ONE WEEK TO TAKE THE PHOTOS; ONE HOUR TO DO THE WRITING EXERCISE (WHILE IT IS NOT NECESSARY TO TAKE A PHOTOGRAPH EVERY DAY, ASSUME THAT THE PHOTOGRAPHY WILL TAKE YOU A TOTAL OF 90 MINUTES.)

Research suggests that finding greater meaning in life helps people cope with stress and improves their overall health and well-being—it's what makes life feel worth living. But finding meaning in life can sometimes feel like an elusive task. In our day-to-day lives, it can be easy to lose sight of the big picture—we tend to focus more on the mundane than the deeply meaningful.

Yet research suggests that there are potential sources of meaning all around us, from the moments of connection we share with others, to the beauty of nature, to the work that we do and the things we create. This exercise helps you bring these meaningful things into focus—literally. By having you photograph, then write about, things that are meaningful to you, it encourages you to pay closer attention to the varied sources of meaning in your life, large and small, and reflect on why they are important to you.

HOW TO DO IT

1. Over the next week, take photographs of things that make your life feel meaningful or full of purpose. These can be people, places, objects, pets, and so on. If you are not able to take photos of these things—like if they're not nearby—you can take photos of souvenirs, reminders, websites, or even other photos. Try to take at least nine photographs.

2. At the end of the week: If you used a phone, upload your photos to a computer. If you used a non-digital camera, have your photos developed.

3. Then, once you have collected your photos, take time to look at and reflect on each one. For each photo, write down a response to the following question: "What does this photo represent, and why is it meaningful?"

PHOTOGRAPH 1: ______

PHOTOGRAPH 2: ______

PHOTOGRAPH 3: ______

PHOTOGRAPH 4: ______

PHOTOGRAPH 5: ______

PHOTOGRAPH 6: ______

PHOTOGRAPH 7: ______

PHOTOGRAPH 8: ______

PHOTOGRAPH 9: ______

WHY IT WORKS

Taking time to recognize and appreciate sources of meaning through photography can help make them more tangible, serving as a reminder of what matters most to you. This greater sense of meaning can, in turn, inspire us to pursue important personal goals and give us a sense of strength and purpose when coping with stressful life events. The use of photography might also benefit people who are more visual than verbal learners—something for therapists, parents, or teachers to keep in mind as they approach conversations about meaning, purpose, and values in life.

REFLECT ON HOW CREATIVITY LEADS TO GRATITUDE.

PURPOSE

HAVING A SENSE OF PURPOSE IS GOOD FOR OUR HEALTH: People with a sense of purpose live longer and may have healthier immune systems.

CERTAIN EXPERIENCES CAN HELP US FIND OUR PURPOSE: They include traveling abroad, spending time in nature, working toward social change, and developing a contemplative practice.

CULTIVATING AWE MIGHT BE ANOTHER WAY TO SHARPEN OUR SENSE OF PURPOSE: Awe connects us to something larger than ourselves, often in the presence of beautiful art, a vast landscape, or a perspective-shifting idea.

PURPOSE PARTICULARLY MATTERS AT WORK: Employees with a sense of purpose at work have higher levels of well-being, and companies with a clear sense of purpose are more likely to have strong financial performances.

"PURPOSE" CAN MEAN MANY THINGS: Having a sense of meaning or purpose is not always the same as being happy.

Research provided by UC Berkeley's Greater Good Science Center.

GOAL VISUALIZATION

TIME REQUIRED: 10 MINUTES DAILY FOR 3 WEEKS

When we face a daunting task, sometimes the hardest part is getting started. To help you overcome that big initial hurdle, this exercise asks you to describe a short-term goal and to visualize the steps you will take to achieve it. In the process, it helps build your confidence that you will be able to reach that goal.

Having confidence in your ability to achieve your goals is a key component of optimism, which research links to greater health and happiness, including lower rates of depression, a better ability to cope with stress, and more relationship satisfaction.

HOW TO DO IT

1. Identify one goal that you would like to achieve in the next day or two and briefly describe it in writing. Make sure that this goal is realistic and not too time-consuming (e.g., "tidy up the hall closet" rather than "clean the entire house top to bottom") and something that is important to you (e.g., "spend quality time with the kids" rather than "learn about the life cycle of the common fly").

2. To help yourself visualize how you will go about accomplishing this goal, describe in writing the steps that you will take to get there.

   For example, if your goal is to tidy up the hall closet, these are the steps that you might take to achieve it:

   a. schedule one hour tonight that you will devote to cleaning

   b. turn off your cell phone/eliminate other distractors

   c. put on some comfortable clothes

   d. turn on some upbeat music

   e. break down the job into sub-tasks: take everything out of the closet, sweep the floor, dust the shelves, get rid of stuff that you don't need anymore, sort the things that you want to keep and put them in boxes, put the boxes back in the closet

   f. remind yourself that it's okay if you don't do everything perfectly or complete the entire task

STEP 1:

STEP 2:

STEP 3:

STEP 4:

STEP 5:

STEP 6:

WHY IT WORKS

This exercise makes goals feel attainable and manageable. When you believe that you will be successful at a goal, it encourages you to work harder toward achieving it—and this greater effort increases the chance that you will actually succeed. Plus, the more you succeed, the more confident you will be about future goals.

Remember, though, not to get down on yourself if you don't succeed right away or perform perfectly. With repeated practice, you may feel greater confidence in your ability to achieve important goals in your life, and this can have a significant impact on your general mood, as regularly completing the goal visualization exercise helps you develop a more optimistic mindset.

BEST POSSIBLE SELF

TIME REQUIRED: 15 MINUTES DAILY FOR 2 WEEKS

Sometimes our goals in life can be elusive. But research suggests that building optimism about the future can motivate people to work toward that desired future and thus make it more likely to become a reality.

This exercise asks you to imagine your life going as well as it possibly could, then write about this best possible future. By doing so, research suggests that you'll not only increase your happiness in the present but also pave the way for sustained happiness down the line.

HOW TO DO IT

Take a moment to imagine your life in the future. What is the best possible life you can imagine? Consider all relevant areas of your life, such as your career, academic work, relationships, hobbies, and/or health. What would happen in these areas of your life in your best possible future?

For the next 15 minutes, write continuously about what you imagine this best possible future to be. Use the instructions below to help guide you through this process.

1. It may be easy for this exercise to lead you to examine how your current life may not match this best possible future. You may be tempted to think about ways in which accomplishing goals has been difficult for you in the past, or about financial/time/social barriers to making these accomplishments happen. For the purpose of this exercise, however, we encourage you to focus on the future—imagine a brighter future in which you are your best self and your circumstances change just enough to make this best possible life happen.

2. This exercise is most useful when it is very specific—if you think about a new job, imagine exactly what you would do, who you would work with, and where it would be. The more specific you are, the more engaged you will be in the exercise and the more you'll get out of it.

3. Be as creative and imaginative as you want, and don't worry about grammar or spelling.

WHY IT WORKS

By thinking about your best possible future self, you can learn about yourself and what you want in life. This way of thinking can help you restructure your priorities in life in order to reach your goals. Additionally, it can help you increase your sense of control over your life by highlighting what you need to do to achieve your dreams.

REFLECT ON HOW PURPOSE LEADS TO GRATITUDE.

GENEROSITY

GENEROSITY CAN MAKE US HAPPIER: Studies find that generosity triggers the release of endorphins in our body, a phenomenon commonly referred to as a "helper's high."

AND HAPPINESS CAN MAKE US MORE GENEROUS: People who feel happy are more likely to be kind to others, creating an upward spiral of happiness and kindness.

GENEROSITY IS LINKED TO LOWER STRESS AND BETTER HEALTH: One study found that people who provided social support to others had lower blood pressure than those who didn't.

GENEROSITY CAN BE CONTAGIOUS: People who come into contact with generous people often follow suit. In fact, research has shown that kindness can spread by three degrees—from person to person to person to person.

GENEROSITY IS A SKILL THAT CAN BE DEVELOPED: Research suggests it's possible to increase your capacity for generosity over time—for instance, by broadening your social networks, actively trying to understand someone else's perspective, or even by meditating.

Research provided by UC Berkeley's Greater Good Science Center.

REMEMBERING CONNECTION

TIME REQUIRED: 10 MINUTES (TRY TO DO THIS PRACTICE AT LEAST ONCE PER WEEK, SELECTING A DIFFERENT EXAMPLE EACH TIME.)

Humans have a strong drive to be kind, but that drive is usually stronger when they feel connected to other people. To help foster that feeling of closeness, this exercise asks you to think about a time when you felt a strong connection to another person and to describe the experience in writing. Research suggests that reflecting on feelings of connection can increase people's motivation to help others, whether by helping a friend or stranger in need, volunteering, or donating money. Helping others can, in turn, increase happiness and improve relationships.

HOW TO DO IT

Try to think of a time when you felt a strong bond with someone in your life. Choose a specific example of an experience you had with this person where you felt especially close and connected to them. This could be a time you had a meaningful conversation, gave or received support, experienced a great loss or success together, or witnessed a historic moment together.

Once you've thought of a specific example, spend a few minutes writing about what happened. In particular, consider the ways in which this experience made you feel close and connected to the other person.

EXAMPLE 1: ____________________

____

____

____

EXAMPLE 2: ____________________

____

____

____

EXAMPLE 3: ____________________

____

____

EXAMPLE 4:

EXAMPLE 5:

EXAMPLE 6:

EXAMPLE 7:

WHY IT WORKS

Feeling connected to others is considered to be a fundamental psychological need. When people feel rejected or alone, they may be more likely to focus on themselves and their own unmet needs rather than attending to the needs of others. When people feel connected and cared for, by contrast, they are better able to expend energy on helping and caring for others.

By reflecting on times when you've felt a strong connection with others, and by striving to cultivate more of these experiences, you are fueling your drive to practice kindness and compassion.

FIND COMMONALITIES

TIME REQUIRED: 15 MINUTES TO GO THROUGH THE STEPS BELOW. TRY TO REPEAT THESE STEPS WITH A DIFFERENT PERSON AT LEAST ONCE PER WEEK

Research suggests that humans have a deeply rooted propensity to be kind and generous, but some obstacles can prevent us from acting on those altruistic impulses. One of the greatest barriers to altruism is that of group difference: We feel much less motivated to help someone if they don't seem to belong to our group or tribe—that is, if they're not a member of our "in-group"—and we may even feel hostile toward members of an "out-group."

But studies have consistently found that who we see as part of our "in-group" can be malleable. That's why one key element of promoting altruism, which involves acting to promote someone else's welfare even at a risk or cost to oneself, is recognizing commonalities with someone else, even if those similarities aren't immediately apparent. This exercise is designed to help expand one's sense of shared identity with others.

HOW TO DO IT

1. Think of a person in your life who seems different from you in every way you can imagine. They might have different interests, different religious or political beliefs, or different life experiences. They may even be someone with whom you have had a personal conflict, or they may belong to a group that has been in conflict with a group to which you belong.

2. Next, make a list of things that you likely have in common with this person. Perhaps you both work for the same company or go to the same school. Maybe you both have children or a significant other. You have probably both had your heart broken at some point, or have lost a loved one. At the broadest level, you both belong to the human species, which means that you share 99.9% of your DNA.

3. Review this list of commonalities. How do they make you see this person in a new light? Instead of simply seeing this person as someone unfamiliar to you, or as a member of an out-group, now try to see this person as an individual, one whose tastes and experiences might overlap with yours in certain ways.

4. Repeat this exercise whenever you meet someone who initially seems different from you, with whom you have a conflict, or who makes you feel uncomfortable.

PERSON 1: ______________________________

PERSON 2: ______________________________

PERSON 3: ______________________________

WHY IT WORKS

Although people generally want and try to be altruistic, they may also feel competitive toward people outside of their "in-group," and the boundaries of their in-group might shrink at times when resources seem scarce or they are fearful for their safety. Reminding people to see the basic humanity that they share with those who might seem different from them can help overcome fear and distrust and promote cooperation. Even small similarities, like recognizing a shared love of sports, can foster a greater sense of kinship across group boundaries. Importantly, recognizing commonalities doesn't mean negating differences, but may in fact help people value differences rather than feeling threatened by them.

RANDOM ACTS OF KINDNESS

TIME REQUIRED: VARIES, DEPENDING ON YOUR ACTS OF KINDNESS—ANYWHERE FROM SEVERAL MINUTES TO SEVERAL HOURS

We all perform acts of kindness at one time or another. These acts may be large or small, and their beneficiaries may not even be aware of them. Yet their effects can be profound—not only on the recipient but on the giver. This exercise asks you to perform five acts of kindness in one day as a way of both promoting kindness in the world and cultivating happiness in yourself and others.

HOW TO DO IT

One day this week, perform five acts of kindness—all five in one day. It doesn't matter if the acts are big or small, but it is more effective if you perform a variety of acts.

The acts do not need to be for the same person—the person doesn't even have to be aware of them. Examples include feeding a stranger's parking meter, donating blood, helping a friend with a chore, or providing a meal to a person in need.

After each act, write down what you did in at least one or two sentences; for more of a happiness boost, also write down how it made you feel.

ACT 1: ______________________________

____

____

____

____

____

ACT 2: ______________________________

____

____

____

____

ACT 3:

ACT 4:

ACT 5:

WHY IT WORKS

Researchers believe this practice makes you feel happier because it makes you think more highly of yourself and become more aware of positive social interactions. It may also increase your kind, helpful—or "pro-social"—attitudes and tendencies toward others. Evidence suggests that variety is key: People who perform the same acts over and over show a downward trajectory in happiness, perhaps because any act starts to feel less special the more it becomes routine.

REFLECT ON HOW GENEROSITY LEADS TO GRATITUDE.

ENERGY

ENERGETIC PEOPLE EXPERIENCE A ZEST FOR LIFE: Zest is part of the character strength of courage, and it means approaching life enthusiastically, as an adventure.

ENERGETIC LIVES LEAD TO POSITIVE RESULTS: People who live life with energy and zest are more satisfied with their lives, find more meaning in life, and experience less anxiety and depression.

TRAUMATIC EVENTS CAN ACTUALLY INCREASE OUR ZEST FOR LIFE: After a struggle, we vow to live life to the fullest rather than watching from the sidelines.

ZESTFUL ENERGY HELPS US ENJOY WORK: People who are zestful and energetic are more likely to see their work as a calling and be satisfied at work.

POSITIVE HABITS INCREASE ENERGY: We can increase our zest and energy by getting enough sleep, engaging in physical activity, cultivating optimism, and spending time with dear friends.

Research provided by UC Berkeley's Greater Good Science Center.

WONDER VIDEO

TIME REQUIRED: 4 MINUTES, PLUS TIME FOR JOURNALING

It's easy to feel bogged down by daily routines and mundane concerns, which can stifle our sense of creativity and wonder. Experiencing awe can reawaken those feelings of inspiration.

Awe is induced by experiences that challenge and expand our typical way of seeing the world, often because we sense that we're in the presence of something greater than ourselves. Research suggests that experiencing awe improves people's satisfaction with life, makes them feel like they have more time, makes them feel less self-conscious, and reduces their focus on trivial concerns.

But in our everyday lives, we might not regularly encounter things that fill us with awe. That's where this practice comes in. It's a way to infuse your day with a dose of wonder even if you can't make it to an inspiring vista or museum.

HOW TO DO IT

Set aside four minutes to watch the video linked in the QR code. Put the video in full-screen mode and try to give it your full attention.

Note that this video is just one example of a visual experience that can elicit awe. There are countless others and being exposed to them can have similar effects. The videos and other stimuli that inspire awe tend to share two key features:

1. They involve a sense of vastness that puts into perspective your own relatively small place in the world. This vastness could be either physical (e.g. a panoramic view from a mountaintop) or psychological (e.g. an exceptionally courageous or heroic act of conscience).

2. They alter the way you understand the world. For instance, they might make your everyday concerns seem less important, or they might expand your beliefs about the reaches of human potential.

JOURNAL ON A MOMENT THAT YOU FELT AWE.

WHY IT WORKS

Taking time out to experience awe can help people break up their routine and challenge themselves to think in new ways. Evoking feelings of awe may be especially helpful when people are feeling bogged down by day-to-day concerns. Research suggests that awe has a way of lifting people outside of their narrower sense of self and connecting them with something larger and more significant. This sense of broader connectedness and purpose can help relieve negative moods and improve happiness.

REFLECT ON HOW ENERGY LEADS TO GRATITUDE.

FOCUS

FOCUS LEADS TO HAPPINESS: A Harvard University study of more than 15,000 people found that people were significantly happier when their minds were focused on what they're doing; they were much less happy when their minds were wandering.

HAPPINESS CAN IMPROVE FOCUS: Research by Barbara Fredrickson of the University of North Carolina suggests that experiencing positive emotions can help people focus on the big picture and avoid getting distracted by minutiae.

FOCUS CAN BE HARD: Almost half of the time, people are thinking about something other than what they're currently doing, ranging from 65 percent of the time while taking a shower to 50 percent while working to 10 percent while having sex.

FOCUS CAN BE SHARPENED (WITH PRACTICE): Research suggests that practicing mindfulness meditation can strengthen attention skills and make people less prone to distraction.

FOCUS LEADS TO SUCCESS: A study of more than 1,000 children in New Zealand found that children who showed stronger powers of concentration and self-control enjoyed better health and wealth and fewer run-ins with the law as adults, regardless of their social class.

FOCUS IS GOOD FOR RELATIONSHIPS: People who are more attuned to other people's facial expressions and body language presented with better social skills and fewer relationship problems.

Research provided by UC Berkeley's Greater Good Science Center.

RAISIN MEDITATION

TIME REQUIRED: 5 MINUTES DAILY FOR AT LEAST A WEEK (EVIDENCE SUGGESTS THAT MINDFULNESS INCREASES THE MORE YOU PRACTICE IT.)

Many of us spend our lives rehashing the past or rushing into the future without pausing to enjoy the present. Distracted from the world around us, our lives might feel only half-lived, as we're too busy to savor—or even notice—everyday pleasures.

Practicing mindfulness can help. Mindfulness helps us tune into what we're sensing and experiencing in the present moment—it's the ability to pay more careful attention to our thoughts, feelings, and bodily sensations without judging them as good or bad. Research suggests that it not only reduces stress but also increases our experience of positive emotions.

One of the most basic and widely used methods for cultivating mindfulness is to focus your attention on each of your senses as you eat a raisin. This simple exercise is often used as an introduction to the practice of mindfulness. In addition to increasing mindfulness generally, the raisin meditation can promote mindful eating and foster a healthier relationship with food. Try it with a single raisin—you might find that it's the most delicious raisin you've ever eaten.

HOW TO DO IT

1. **HOLDING:** First, take a raisin and hold it in the palm of your hand or between your finger and thumb.

2. **SEEING:** Take time to really focus on it; gaze at the raisin with care and full attention—imagine that you've just dropped in from Mars and have never seen an object like this before in your life. Let your eyes explore every part of it, examining the places where light shines on it, the darker hollows, the folds and ridges, and any asymmetries or unique features.

3. **TOUCHING:** Feel the raisin between your fingers, exploring its texture. Do this with your eyes closed if that enhances your sense of touch.

4. **SMELLING:** Hold the raisin beneath your nose. With each inhalation, take in any smell, aroma, or fragrance. As you do this, notice anything interesting that may be happening in your mouth or stomach.

5. **Placing:** Now, slowly bring the raisin up to your lips, noticing how your hand and arm know exactly how and where to position it. Gently place the raisin in your mouth; without chewing, notice how it gets into your mouth in the first place. Spend a few moments focusing on the sensations of having it in your mouth.

6. **Tasting:** When you are ready, prepare to chew the raisin, noticing how and where it needs to be positioned in your mouth for chewing. Then, very consciously, take one or two bites into it and notice what happens in the aftermath, experiencing any waves of taste that emanate from it as you continue chewing. Without swallowing yet, notice the bare sensations of taste and texture in your mouth and how these may change, moment by moment. Pay attention to any changes in the rainsin.

7. **Swallowing:** When you feel ready to swallow the raisin, see if you can first detect the reflex to swallow as it comes up, so that even this is experienced consciously before you actually swallow the raisin.

8. **Following:** Finally, see if you can feel what is left of the raisin moving down into your stomach, and sense how your body as a whole is feeling after you have completed this exercise.

WHY IT WORKS

By increasing awareness of internal mental and physical states, mindfulness can help people gain a greater sense of control over their thoughts, feelings, and behaviors in the present moment. Paying closer attention to the sensations of eating can increase our enjoyment of our food and deepen our appreciation for the opportunity to satisfy our hunger. Mindfulness can also help people become more attuned to hunger and fullness signals and therefore avoid overeating or "emotional eating." In the words of mindfulness expert Jon Kabat-Zinn, "When we taste with attention, even the simplest foods provide a universe of sensory experience."

BODY SCAN

TIME REQUIRED: 20-45 MINUTES, THREE TO SIX DAYS PER WEEK FOR FOUR WEEKS. RESEARCH SUGGESTS THAT PEOPLE WHO PRACTICE THE BODY SCAN FOR LONGER REAP MORE BENEFITS FROM THIS PRACTICE.

This exercise asks you to systematically focus your attention on different parts of your body, from your feet to the muscles in your face. It is designed to help you develop a mindful awareness of your bodily sensations and to relieve tension wherever it is found. Research suggests that this mindfulness practice can help reduce stress, improve well-being, and decrease aches and pains.

HOW TO DO IT

The body scan can be performed while lying down, sitting, or in almost any other posture. The steps below are a guided meditation designed to be done while sitting. You can listen to audio of this three-minute guided meditation, produced by UCLA's Mindful Awareness Research Center (MARC), linked in this QR code.

Especially for those new to the body scan, performing this practice with the audio is recommended. However, you can also use the script below for guidance for yourself or to lead this practice for others.

- Begin by bringing your attention into your body.
- Close your eyes if that's comfortable for you.
- Notice your body seated wherever you're seated, feeling the weight of your body on the chair or on the floor.
- Take a few deep breaths.
- As you take a deep breath, bring in more oxygen, enlivening the body. As you exhale, have a sense of relaxing more deeply.
- Notice your feet on the floor—the sensation of your feet touching the floor—the weight, pressure, vibration, and heat.
- Notice your legs against the chair, pressure, pulsing, heaviness, and lightness.
- Notice your back against the chair.

- Bring your attention into your stomach area. If your stomach is tense or tight, let it soften. Take a breath.
- Notice your hands. Are your hands tense or tight? See if you can allow them to soften.
- Notice your arms. Feel any sensation in your arms. Let your shoulders be soft.
- Notice your neck and throat. Let them be soft. Relax.
- Soften your jaw. Let your face and facial muscles be soft.
- Then, notice your whole body present. Take one more breath.
- Be aware of your whole body as best you can. Take a breath. Then, when you're ready, you can open your eyes.

You can also listen to a 45-minute version of the Body Scan that the UC San Diego Center for Mindfulness uses in its trainings in Mindfulness-Based Stress Reduction through this QR code.

WHY IT WORKS

The body scan provides a rare opportunity for us to experience our body as it is, without judging or trying to change it. It may allow us to notice and release a source of tension we weren't aware of before, such as a hunched back or clenched jaw muscles. Or, it may draw our attention to a source of pain and discomfort. Our feelings of resistance and anger toward pain often only serve to increase that pain and to increase the distress associated with it. According to research, by simply noticing the pain we're experiencing—without trying to change it—we may actually feel some relief.

The body scan is designed to counteract these negative feelings toward our bodies. This practice may also increase our general attunement to our physical needs and sensations, which can in turn help us take better care of our bodies and make healthier decisions about eating, sleep, and exercise.

REFLECT ON HOW FOCUS LEADS TO GRATITUDE.

COURAGE

COURAGE INVOLVES RISK—AND PERHAPS FEAR: When scientists define courage, they emphasize that it means voluntarily putting yourself at risk to achieve a positive goal. Some—but not all—experts believe that courage involves confronting and overcoming fear, rather than an absence of fear.

COURAGE IS COMMON AMONG KIDS: In one study, over 70 percent of children between the ages of 8 and 13 reported they had performed at least one courageous act.

COURAGEOUS KIDS ARE OUTGOING: They score higher on questionnaires that measure extraversion and lower on measures of anxiety.

IT'S POSSIBLE TO BUILD COURAGE: Research suggests we can overcome fears—not by avoiding them, but by gradually and repeatedly exposing ourselves to what we fear.

Research provided by UC Berkeley's Greater Good Science Center.

OVERCOMING FEAR

TIME REQUIRED: VARIES. BE MINDFUL AND PATIENT WITH YOURSELF AS YOU GO THROUGH THIS PROCESS.

Some types of fear—like the fear that stops you from running into a busy street—are useful and necessary. But other types of fear are less rational and more likely to hold you back in life. Fear of public speaking, fear of flying, fear of heights—these are some of the more common ones.

To cope, you may avoid the situations that elicit these fears, or you may try, often unsuccessfully, to counter your fear with reason—for example, by reminding yourself of the very low likelihood of a plane crash.

Research suggests that a more effective way to combat fear is to do the thing you least want to do—face your fear head on—but do it one step at a time, in a healthy and safe way. This strategy can help re-train your brain to develop a more positive association with whatever has been triggering your fear. Confronting your fears head-on can also increase your self-confidence as you show yourself that you're capable of doing what might once have seemed impossible. While acting based on fear limits you, facing your fears can be liberating and transformative.

HOW TO DO IT

Note: The following guidelines are geared toward addressing mild, everyday fears. Fears related to serious mental illnesses such as post-traumatic stress disorder, obsessive-compulsive disorder, and social anxiety disorder should be addressed with the help of a mental health professional.

Sometimes, one or two scary experiences can cause us to fear things that we don't rationally need to fear; other fears aren't based on first-hand experience at all. Either way, overcoming these fears often requires that we develop a more positive—or at least less negative—association with the thing that we fear. Here's how:

1. START WITH SMALL DOSES. The first step is to expose yourself to small doses of the fear-inducing activity in a safe context. For example, if public speaking makes you nervous, you could start by seeking out a low-pressure speaking opportunity with a small, supportive audience in a setting where you don't have to worry about being perfectly articulate—perhaps giving a toast at a friend's

birthday party. Or, if you'd like to learn to rock climb but are afraid of heights, you could start by spending time assisting other climbers.

2. **Repeat the activity until you start to feel the fear dissipate.** Over time, repeated exposure to a safe, non-harmful version of whatever made you afraid can reduce the negative association and replace it with a neutral or positive association. For example, repeatedly seeing other people climb without falling may begin to overwrite your negative association with heights. And the more you fly and land safely, the less dangerous flying is likely to feel.

3. **Gradually increase the challenge.** After you begin to feel more comfortable with small doses, try taking it up a notch. For example, you could go from watching others climb to climbing a short distance yourself. Or, you could volunteer to present the results of a team project to coworkers or fellow students. From here, you can continue to incrementally ratchet up the challenge until you reach your goal, whether that's to scale Mt. Everest, give a talk in front of hundreds of people, or fly to a new continent.

Your fear may never be fully extinguished, but hopefully, it will hold less power over you and no longer prevent you from achieving important goals and enjoying your life. In the words of Mark Twain, "Courage is not the absence of fear. It is acting in spite of it."

WHY IT WORKS

Fear may be natural, but it's not always helpful. Sometimes, our brains mistakenly learn to send fear signals even when there is no real danger, perhaps based on one or two bad experiences. Gradually and repeatedly exposing ourselves to the activities we fear most can help teach our brains that these activities are not in fact dangerous—and may actually be very rewarding.

REFLECT ON HOW COURAGE LEADS TO GRATITUDE.

GRATITUDE

GRATEFUL PEOPLE ARE HAPPIER: Research by Sonja Lyubomirsky of the University of California, Riverside, and others has found that grateful people experience more optimism, joy, enthusiasm, and other positive emotions and that they have a deeper appreciation for life's simple pleasures.

GRATITUDE IS GOOD FOR OUR HEALTH: Studies link gratitude to a stronger immune system, lower blood pressure, better sleep quality, reduced risk of heart disease, and better kidney function.

GRATITUDE IMPROVES OUR RELATIONSHIPS: When someone feels grateful for their romantic partner on one day, *both* partners feel more satisfied with their relationship on the next. Expressing gratitude also makes people feel closer to their friends.

GRATITUDE IS GOOD FOR KIDS: Grateful teens are more satisfied with their lives, more engaged at school, have higher grades, and are less materialistic.

GRATITUDE IS A SKILL: People can increase their level of gratitude—and enjoy the benefits—through practice, such as by keeping a gratitude journal.

GRATITUDE MOTIVATES US TO "PAY IT FORWARD": In one study, people who benefited from an act of kindness later spent significantly more time helping others than those who did not.

GRATITUDE MAKES US SMARTER IN OUR SPENDING: People who feel grateful show stronger self-control and are better at delaying gratification, rather than making impulsive, short-sighted spending decisions.

Research provided by UC Berkeley's Greater Good Science Center.

GRATITUDE LETTER

TIME REQUIRED: AT LEAST 15 MINUTES FOR WRITING THE LETTER, AT LEAST 30 MINUTES FOR THE VISIT

Feeling gratitude can improve health and happiness; expressing gratitude also strengthens relationships. Yet sometimes, expressions of thanks can be fleeting and superficial. This exercise encourages you to express gratitude in a thoughtful, deliberate way by writing—and, ideally, delivering—a letter of gratitude to a person you have never properly thanked.

HOW TO DO IT

Call to mind someone who did something for you for which you are extremely grateful but to whom you never expressed your deep gratitude. This could be a relative, friend, teacher, or colleague. Try to pick someone who is still alive and could meet you face-to-face in the next week. It may be most helpful to select a person or act that you haven't thought about for a while—something that isn't always on your mind.

Now, write a letter to one of these people, guided by the following steps.

1. Write as though you are addressing this person directly: ("Dear ___________").
2. Don't worry about perfect grammar or spelling.
3. Describe in specific terms what this person did, why you are grateful to this person, and how this person's behavior affected your life. Try to be as concrete as possible.
4. Describe what you are doing in your life now and how you often remember their efforts.
5. Try to keep your letter to roughly one page (~300 words).

USE THIS SPACE TO HELP YOU DRAFT YOUR LETTER.

Next, you should try if at all possible to deliver your letter in person, following these steps:

1. Plan a visit with the recipient. Let that person know you'd like to see them and have something special to share, but don't reveal the exact purpose of the meeting.
2. When you meet, let the person know that you are grateful to them and would like to read a letter expressing your gratitude; ask that they refrain from interrupting until you're done.
3. Take your time reading the letter. While you read, pay attention to their reaction as well as your own.
4. After you have read the letter, be receptive to their reaction and discuss your feelings together.
5. Remember to give the letter to the person when you leave.

If physical distance keeps you from making a visit, you may choose to arrange a phone or video chat.

WHY IT WORKS

The letter affirms positive things in your life and reminds you how others have cared for you—life seems less bleak and lonely if someone has taken such a supportive interest in us. Visiting the giver allows you to strengthen your connection with her and remember how others value you as an individual.

GRATITUDE JOURNAL

TIME REQUIRED: 15 MINUTES DAILY, AT LEAST ONCE PER WEEK FOR AT LEAST TWO WEEKS (STUDIES SUGGEST THAT WRITING IN A GRATITUDE JOURNAL THREE TIMES PER WEEK MIGHT ACTUALLY HAVE A GREATER IMPACT ON OUR HAPPINESS THAN JOURNALING EVERY DAY.)

It's easy to take the good things and people in our lives for granted, but research suggests that consciously giving thanks for them can have profound effects on our well-being and relationships. This exercise helps you develop a greater appreciation for the good in your life. In fact, people who routinely express gratitude enjoy better health and greater happiness.

HOW TO DO IT

There's no wrong way to keep a gratitude journal, but here are some general instructions as you get started.

Write down up to five things for which you feel grateful on the following two pages. Keeping a physical record is important—don't just do this exercise in your head. The things you list can be relatively small in importance ("The tasty sandwich I had for lunch today.") or relatively large ("My sister gave birth to a healthy baby boy."). The goal of the exercise is to remember a good event, experience, person, or thing in your life—then enjoy the good emotions that come with it.

As you write, here are nine important tips:

1. **BE AS SPECIFIC AS POSSIBLE—SPECIFICITY IS KEY TO FOSTERING GRATITUDE.** "I'm grateful that my co-workers brought me soup when I was sick on Tuesday" will be more effective than "I'm grateful for my co-workers."
2. **GO FOR DEPTH OVER BREADTH.** Elaborating in detail about a particular person or thing for which you're grateful carries more benefits than a superficial list of many things.
3. **GET PERSONAL.** Focusing on people to whom you are grateful has more of an impact than focusing on things for which you are grateful.
4. **TRY SUBTRACTION, NOT JUST ADDITION.** Consider what your life would be like without certain people or things, rather than just tallying up all the good stuff. Be grateful for the negative outcomes

you avoided, escaped, prevented, or turned into something positive—try not to take that good fortune for granted.

5. **SEE GOOD THINGS AS "GIFTS."** Thinking of the good things in your life as gifts helps you avoid taking them for granted. Try to relish and savor the gifts you've received.
6. **SAVOR SURPRISES.** Try to record events that were unexpected or surprising, as these tend to elicit stronger levels of gratitude.
7. **IF YOU REPEAT, REVISE.** Writing repeatedly about some of the same people and things is okay, but focus on a different aspect each time.
8. **WRITE REGULARLY.** Whether you write every other day or once a week, commit to a regular time to journal and honor that commitment.
9. **DON'T OVERDO IT.** Evidence suggests writing occasionally (1-3 times per week) is more beneficial than daily journaling. That might be because as we adapt to positive events, we can become numb to them—that's why it helps to savor surprises.

WHY IT WORKS

While it's important to analyze and learn from negative events, sometimes we can think too much about what goes wrong and not enough about what goes right in our lives. A gratitude journal forces us to pay attention to the good things in life we might otherwise take for granted. In that way, we start to become more attuned to the everyday sources of pleasure around us—and the emotional tone of our life can shift in profound ways. What's more, actually writing about these events is key: Research suggests translating thoughts into concrete language makes us more aware of them, deepening their emotional impact.

HAPPINESS

HAPPINESS DOESN'T SIMPLY MEAN FEELING JOYFUL ALL THE TIME: Leading happiness researcher Sonja Lyubomirsky writes that happiness involves feeling positive emotions along with "a sense that one's life is good, meaningful, and worthwhile." Some research even suggests that trying to be happy all the time is a recipe for unhappiness.

HAPPY PEOPLE ENJOY BETTER HEALTH: Studies suggest that happiness reduces your risk of heart disease, strengthens your immune system, and may ultimately add years to your life.

HAPPY PEOPLE ARE MORE GENEROUS: Studies suggest that feeling happy makes you kind—and kindness makes you happy—an upward spiral of happiness and goodness.

HAPPY PEOPLE "PRIORITIZE POSITIVITY": They deliberately organize their day-to-day lives to include situations that naturally give rise to positive emotions.

HAPPINESS IS GOOD FOR OUR RELATIONSHIPS: Happy people are more likely to get married, have fulfilling marriages, and have more friends.

MORE MONEY DOESN'T BRING MORE HAPPINESS: Happy people generally do make more money and are more productive at work, but research suggests that once people earn roughly $75,000 per year, more pay doesn't bring more happiness.

Research provided by UC Berkeley's Greater Good Science Center.

THREE GOOD THINGS

TIME REQUIRED: 10 MINUTES DAILY FOR AT LEAST ONE WEEK

In our day-to-day lives, it's easy to get caught up in the things that go wrong and feel like we're living under our own private rain cloud; at the same time, we tend to adapt to the good things and people in our lives, taking them for granted. As a result, we often overlook everyday beauty and goodness—a kind gesture from a stranger, say, or the warmth of our heater on a chilly morning. In the process, we frequently miss opportunities for happiness and connection.

This practice guards against those tendencies. By remembering and listing three positive things that have happened in your day—and considering what caused them—you tune into the sources of goodness in your life. It's a habit that can change the emotional tone of your life, replacing feelings of disappointment or entitlement with those of gratitude—which may be why this practice is associated with significant increases in happiness.

HOW TO DO IT

Each day for at least one week, write down three things that went well for you, and provide an explanation for why they went well. It is important to create a physical record of your experiences by writing them down; it is not enough to do this exercise only in your head. The things you write can be relatively small in importance (e.g., "my coworker made me coffee today") or relatively large (e.g., "I earned a big promotion"). To make this exercise part of a daily routine, some find that writing before bed is helpful.

As you write, follow these instructions:

1. Give the event a title (e.g., "coworker complimented my work on a project")
2. Write down exactly what happened in as much detail as possible, including what you did or said and, if others were involved, what they did or said.
3. Include how this event made you feel at the time and how it made you feel later (including now, as you remember it).
4. Explain what you think caused this event—why it came to pass.

Use whatever writing style you please, and do not worry about perfect grammar and spelling. Use as much detail as you'd like.

If you find yourself focusing on negative feelings, refocus your mind on the good event and the positive feelings that came with it. This can take effort, but it gets easier with practice and can make a real difference in how you feel.

DAY 1

EVENT 1

EVENT 2

EVENT 3

DAY 2

EVENT 1

EVENT 2

EVENT 3

DAY 3

EVENT 1

EVENT 2

EVENT 3

DAY 4

EVENT 1

EVENT 2

EVENT 3

DAY 5

EVENT 1

EVENT 2

EVENT 3

DAY 6

EVENT 1

EVENT 2

EVENT 3

DAY 7

EVENT 1

EVENT 2

EVENT 3

WHY IT WORKS

By giving you the space to focus on the positive, this practice teaches you to notice, remember, and savor the better things in life. It may prompt you to pay closer attention to positive events down the road and engage in them more fully—both in the moment and later on, when you can reminisce and share your experiences with others. Reflecting on the cause of the event may help attune you to the deeper sources of goodness in your life, fostering a mindset of gratitude.

POSITIVE EVENTS

TIME REQUIRED: THE BETTER PART OF ONE DAY

One of the most direct ways to increase our happiness is to do more of the things that make us happy. But when life gets busy, we don't always remember to make time for enjoyable activities. Intentionally scheduling a variety of these activities into your day can help you overcome this barrier to happiness.

This exercise prompts you to engage in a variety of activities you associate with happiness and reflect on how they make you feel. Different kinds of activities bring different kinds of satisfaction, all of which contribute uniquely to happiness. Research suggests that including variety and novelty in daily activities is an important component of happiness, so trying a number of different activities can help prevent you from doing any one activity so habitually that it ceases to bring you joy.

HOW TO DO IT

This exercise is best completed on a day (or two) when you have a lot of free time, such as on a weekend. Step two may require some advanced planning with others. On the morning of your free day when you first wake up, review the following instructions and make a plan for the day. Write your plan for each activity on the lines provided.

1. Choose an activity that you enjoy doing alone, such as reading, listening to music, watching a TV show, or meditating. Set aside some time during the day to complete this activity.
2. Choose an activity that you enjoy doing with others, such as going out for coffee, going for a bike ride, or watching a movie. Set aside some time during the day to complete this activity.
3. Choose an activity that you consider personally important and meaningful, such as helping a neighbor, calling to check in on a friend who is sick, or volunteering for a local charitable organization.
4. At the end of the day, record what occurred during and after each of your three activities. What did you do, and how did it make you feel? Did different activities make you feel different kinds of happiness? What feelings or associations linger with you now, after you have completed all of the activities?

Activity 1: ______________________________

Activity 2: ______________________________

Activity 3: ______________________________

WHY IT WORKS

Engaging in a range of activities can prevent "habituation," which is the tendency to get used to things that we do regularly and thus derive less satisfaction from them. What's more, research suggests that combining activities that are related to different kinds of happiness (e.g., pleasure-based happiness vs. meaning-based happiness) can promote greater overall happiness. Reflecting on enjoyable activities at the end of the day can make us more likely to store those experiences as memories and derive pleasure from them later on.

REFLECT ON HOW HAPPINESS LEADS TO GRATITUDE.

PATIENCE

PATIENCE IS A GREAT VIRTUE: Patience is the ability to remain calm in the face of frustration or adversity. It's a quiet virtue encouraged by many world religions and philosophers.

PATIENCE IS AN ATTITUDE: Patience is one of the "nine attitudes of mindfulness" described by mindfulness teacher, researcher, and best-selling author Jon Kabat-Zinn. While impatience betrays a desire to be done with the here-and-now and move on to the future, patience means we're focused on the way things are in the present.

PATIENCE IS GOOD FOR OUR HEALTH: Patient people experience fewer symptoms of illness, such as headaches and ulcers. In contrast, people who exhibit impatience and irritability tend to have more health complaints and worse sleep quality.

PATIENCE MAKES PEOPLE MORE "HUMAN": Patient people are less likely to experience negative emotions and depression. They tend to be more equitable and forgiving, have more self-control, and show more empathic concern for others.

WE CAN INCREASE PATIENCE WITH MINDFULNESS AND GRATITUDE: Children who did a six-month mindfulness program in school became less impulsive and more willing to wait for a reward. Adults who are feeling grateful are also better at delaying gratification.

Research provided by UC Berkeley's Greater Good Science Center.

MINDFUL BREATHING

TIME REQUIRED: 15 MINUTES DAILY FOR AT LEAST ONE WEEK (EVIDENCE SUGGESTS THAT MINDFULNESS INCREASES THE MORE YOU PRACTICE IT.)

Stress, anger, and anxiety can impair not only our health but our judgement and skills of attention. Fortunately, research suggests an effective way to deal with these difficulty feelings: practicing "mindfulness," the ability to pay careful attention to what you're thinking, feeling, and sensing in the present moment without judging those thoughts and feelings as good or bad. Countless studies link mindfulness to better health, lower anxiety, and greater resilience to stress.

But how do you cultivate mindfulness? A basic method is to focus your attention on your own breathing—a practice called, quite simply, "mindful breathing." After setting aside time to practice mindful breath, you should find it easier to focus your attention on your breath in your daily life—an important skill to help you deal with stress, anxiety, and negative emotions, cool yourself down when your temper flares up, and sharpen your skills of concentration.

HOW TO DO IT

The most basic way to do mindful breathing is to focus your attention on your breath, following the inhale and the exhale. You can do this while standing, but ideally, you'll be sitting or even lying in a comfortable position. Your eyes may be open or closed, but you may find it easier to maintain your focus if you close your eyes. It helps to set aside a designated time for this exercise, but it also helps to practice it when you're feeling particularly stressed or anxious. Experts believe that making a regular practice of mindful breathing makes it easier to breathe mindfully in difficult situations.

Sometimes, especially when trying to calm yourself in a stressful moment, it might help to start by taking an exaggerated breath: inhale deeply through your nostrils (3 seconds), hold your breath (2 seconds), and exhale slowly through your mouth (4 seconds). Otherwise, simply observe each breath without trying to adjust it; it may help to focus on the rise and fall of your chest or the sensation of air flowing through your nostrils. As you breathe, you may find that your mind wanders, distracted by thoughts or bodily sensations. That's okay. Just notice that it is happening and gently bring your attention back to your breath.

To provide even more structure, and help you lead this practice for others, below are steps for a short guided meditation. You can listen to audio of this guided meditation, produced by UCLA's Mindful Awareness Research Center (MARC), in the QR code here.

1. Find a relaxed, comfortable position. You could be seated on a chair or on the floor on a cushion. Keep your back upright, but not too tight. Rest your hands wherever they're comfortable. Rest your tongue on the roof of your mouth or wherever it's comfortable.
2. Notice and relax your body. Try to notice the shape of your body and its weight. Let yourself relax and become curious about your body seated here—the sensations it experiences and its connection with the floor or the chair. Relax any areas of tightness or tension. Just breathe.
3. Tune into your breath. Feel the natural flow of breath—in, out. You don't need to do anything special with your breath. Not long, not short, just natural. Notice where you feel your breath in your body. It might be in your abdomen. It might be in your chest, throat, or nostrils. See if you can feel the sensations of breath, one breath at a time. When one breath ends, the next breath begins.
4. As you do this, your mind may start to wander. You may start thinking about other things. If this happens, it is not a problem. It's very natural. Just notice that your mind has wandered. You can say "thinking" or "wandering" softly to yourself. Then gently redirect your attention right back to the breathing.
5. Stay here for five to seven minutes. Notice your breath in silence. From time to time, you'll get lost in thought; when you do, return to your breath.
6. After a few minutes, once again notice your body—your whole body—seated here. Let yourself relax even more deeply, and then offer yourself some appreciation for doing this practice today.

WHY IT WORKS

Mindfulness gives people distance from their thoughts and feelings, which can help them tolerate and work through unpleasant feelings rather than becoming overwhelmed by them. Mindful breathing in particular is helpful because it gives people an anchor—their breath—on which they can focus when they find themselves carried away by a stressful thought. Mindful breathing also helps people stay "present" in the moment, rather than being distracted by regrets about the past or worries about the future.

ACTIVE LISTENING

TIME REQUIRED: AT LEAST 10 MINUTES (TRY TO MAKE TIME FOR THIS PRACTICE AT LEAST ONCE PER WEEK.)

Often we'll listen to a conversation partner without really hearing them. In the process, we miss opportunities to connect with that person—and even risk making them feel neglected, disrespected, or resentful.

This exercise helps you express active interest in what the other person has to say and make them feel heard—a way to foster empathy and connection. This technique is especially well-suited for difficult conversations (such as arguments with a spouse) and for expressing support. Research suggests that using this technique can help others feel more understood and improve relationship satisfaction.

HOW TO DO IT

Find a quiet place where you can talk with a conversation partner without interruption or distraction. Invite them to share what's on their mind. As they do so, try to follow the steps below. You don't need to cover every step, but the more you cover, the more effective this practice is likely to be.

1. PARAPHRASE. Once the other person has finished expressing a thought, paraphrase what they said to make sure you understand and to show that you are paying attention. Helpful ways to paraphrase include "What I hear you saying is..." "It sounds like..." and "If I understand you right..."
2. ASK QUESTIONS. When appropriate, ask questions to encourage the other person to elaborate on their thoughts and feelings. Avoid jumping to conclusions about what the other person means. Instead, ask questions to clarify their meaning, such as, "When you say_______, do you mean_______?"
3. EXPRESS EMPATHY. If the other person voices negative feelings, strive to validate these feelings rather than questioning or defending against them. For example, if the speaker expresses frustration, try to consider why they feel that way, regardless of whether you think that feeling is justified or whether you would feel that way yourself were you in the same position. You might respond, "I can sense that you're feeling frustrated," and even "I can understand how that situation could cause frustration."

4. **USE ENGAGED BODY LANGUAGE.** Show that you are engaged and interested by making eye contact, nodding, facing the other person, and maintaining an open and relaxed body posture. Avoid distractions in your environment, such as checking your phone. Be mindful of your facial expressions: Avoid expressions that might communicate disapproval or disgust.
5. **AVOID JUDGMENT.** Your goal is to understand the other person's perspective and accept it for what it is, even if you disagree with it. Try to avoid interrupting with counterarguments or mentally preparing a rebuttal while the other person is speaking.
6. **AVOID GIVING ADVICE.** Problem-solving is likely to be more effective after both conversation partners understand one another's perspectives and feel heard. Moving too quickly into advice-giving can be counterproductive.
7. **TAKE TURNS.** After the other person has had a chance to speak and you have engaged in the active listening steps above, ask if it's okay for you to share your perspective. When sharing your perspective, express yourself as clearly as possible using "I" statements (e.g., "I feel overwhelmed when you don't help out around the house"). It may also be helpful, when relevant, to express empathy for the other person's perspective (e.g., "I know you've been very busy lately and don't mean to leave me hanging").

WHY IT WORKS

Active listening helps listeners better understand others' perspectives and helps speakers feel more understood and less threatened. This technique can prevent miscommunication and spare hurt feelings on both sides. By improving communication and preventing arguments from escalating, active listening can make relationships more enduring and satisfying. Practicing active listening with someone close to you can also help you listen better when interacting with other people in your life, such as students, coworkers, or roommates.

REFLECT ON HOW PATIENCE LEADS TO GRATITUDE.

LOVE

LOVE CAN LAST: Scientists disagree with the cynical view that love fades over time. One study found that a significant proportion of long-married couples still reported being intensely in love.

LOVE IS GOOD FOR YOUR BODY: Feeling "in love" boosts the body's level of oxytocin, also known as "the love hormone," which makes you experience a wave of contentment and calms the body.

LOVE IS BLIND—AND THAT CAN BE A GOOD THING: Studies suggest that maintaining "positive illusions" about your partner, such as believing they are exceptionally smart or good looking, is associated with greater relationship satisfaction for both of you.

LOVE IS STRENGTHENED BY GRATITUDE: Consciously feeling and expressing thanks for your partner makes both of you feel closer to one another and more committed to the relationship.

LOVE REDUCES PAIN: People in painful situations report feeling less pain when they look at photographs of their loved ones.

Research provided by UC Berkeley's Greater Good Science Center.

MENTAL SUBTRACTION

TIME REQUIRED: 15 MINUTES (TRY TO MAKE TIME TO DO THIS PRACTICE ONCE PER WEEK, FOCUSING ON A DIFFERENT PERSON EACH WEEK. IT MIGHT HELP TO DO THIS PRACTICE AT THE SAME TIME EACH WEEK—BEFORE BED EACH SUNDAY EVENING, SAY, OR AT LUNCH EVERY FRIDAY.)

It's easy to take the important people in our lives for granted, but research suggests that if we experience and express gratitude for them, our relationships will be stronger and our lives will be happier. This exercise is designed to elicit feelings of gratitude for one of these people—such as a romantic partner or close friend—by asking you to think about what your life might have been like had you never met them. By getting a taste of their absence, you should be able to appreciate their presence in your life more deeply without actually having to lose them.

HOW TO DO IT

1. Take a moment to think about an important relationship in your life, such as a romantic relationship or close friendship.
2. Think back to where and how you met this person.
3. Consider the ways in which you might never have met this person and never formed a friendship or relationship—for example, if you hadn't decided to go to a certain party, taken a certain class, or moved to a certain city.
4. Write down all the possible events and decisions—large and small—that could have gone differently and prevented you from meeting this person.
5. Imagine what your life would be like now if events had unfolded differently and you had never met this person. Bring to mind some of the joys and benefits you have enjoyed as a result of this relationship—and consider how you would feel if you were denied all of them.
6. Shift your focus to remind yourself that you did meet this person and reflect upon the benefits this relationship has brought you. Now that you have considered how things might have turned out differently, appreciate that these benefits were not inevitable in your life. Allow yourself to feel grateful that things happened as they did and that this person is now in your life.

Person 1:

Person 2:

Person 3:

Person 4:

Person 5:

WHY IT WORKS

Mental subtraction counteracts our tendency to take positive events in our lives—such as meeting a great person—as givens. When we consider the circumstances that led to a chance encounter, we may be surprised by how unlikely that meeting actually was, and how lucky we were that it happened as it did. While it can be painful to think about not having met someone close to us, this scenario provides a negative contrast against which our current situation can be favorably compared.

AVOIDING THE FOUR HORSEMEN

TIME REQUIRED: 10 MINUTES TO COMPLETE THE FIRST THREE STEPS; THE AMOUNT OF TIME TO COMPLETE THE REST WILL VARY. (TRY TO GO THROUGH THIS EXERCISE AT LEAST ONCE PER MONTH. AFTER EVALUATING YOUR CLASSROOM, OFFICE, OR A ROOM IN YOUR HOME, NEXT MONTH CONSIDER ANOTHER ROOM OR ENVIRONMENT OVER WHICH YOU HAVE CONTROL.)

Research suggests that humans have a strong propensity for kindness and generosity and that kindness improves the health and happiness of both the giver and the receiver. But we don't always act on our altruistic instincts.

Fortunately, studies have identified ways to elicit people's deeply rooted propensities for kindness. One of the most effective is to evoke a sense of connectedness among people. Research suggests that even subtle reminders of connection, operating below the conscious level, can lead to concrete, measurable increases in altruistic behavior. This exercise walks you through the process of considering how you can add reminders of social connection to your home, office, or classroom.

HOW TO DO IT

1. Take a moment to look around your home, office, or classroom. What kinds of objects, words, and images surround you?
2. Count how many of these objects, words, and images are related to social connectedness. This could include pictures of people interacting, words like "community," "together," or "friendship," or even two stuffed animals facing one another on a shelf.
3. Notice whether there are any empty walls or shelves where you could add new objects related to connectedness, or places where you could replace existing objects.
4. Next time you're out shopping, looking through your belongings, or (for parents or teachers) developing an art project for your children or students, see if you can find objects that evoke connection, even in subtle ways, and use them to fill these empty places or to replace existing objects.
5. Finally, consider how the furniture in this room is arranged. Are chairs facing toward or away from each other? Are there common spaces that are conducive to social interaction? Rearranging the layout of your home, office, or classroom can also help to promote feelings of connectedness.

WHY IT WORKS

Although people generally want to be altruistic, we don't always act that way. This is sometimes simply because we are busy and distracted by other things, like personal problems or approaching deadlines. By creating reminders of social connection in your home, office, or classroom, you disrupt this self-focused tendency and reorient your attention to focus on other people. When we feel connected to others, we are more likely to want to help them—perhaps because, throughout humans' evolutionary history, caring for those close to us was essential to the survival of our species.

REFLECT ON HOW LOVE LEADS TO GRATITUDE.

MINDFULNESS

MINDFULNESS IS ABOUT MORE THAN MEDITATION: It means paying attention to whatever you're thinking, feeling, or sensing in a given moment without judging those thoughts or feelings as "good" or "bad."

MINDFULNESS IS GOOD FOR OUR BODIES: A seminal study found that after just eight weeks of training, practicing mindfulness boosted the immune system's ability to fight off illness.

MINDFULNESS IS GOOD FOR OUR MINDS: Research suggests that mindfulness increases positive emotions while reducing negative emotions and stress. At least one study has found that it may be as effective as antidepressants in fighting depression.

MINDFULNESS CHANGES OUR BRAINS: A Harvard Medical School study found that eight weeks of mindfulness practice increased the density of gray matter in regions of the brain linked to learning, memory, emotion regulation, and empathy.

MINDFULNESS INCREASES KINDNESS AND COMPASSION: Research suggests it makes us more likely to help someone in need and increases activity in neural networks involved in understanding the suffering of others.

MINDFULNESS HELPS US FOCUS: Studies suggest mindfulness helps us tune out distractions and improves our memory and attention skills.

MINDFULNESS HELPS SCHOOLS: Dozens of studies have suggested that mindfulness could be used to improve students' grades, social skills, and behavior and that it may help reduce stress and burnout among teachers.

Research provided by UC Berkeley's Greater Good Science Center.

SELF-COMPASSION BREAK

TIME REQUIRED: 5 MINUTES (WHILE IT MAY BE CHALLENGING TO DO THIS PRACTICE EVERY TIME YOU FACE A STRESSFUL SITUATION, AN INITIAL GOAL COULD BE TO TRY IT AT LEAST ONCE PER WEEK.)

Difficult situations become even harder when we beat ourselves up over them, interpreting them as a sign that we're less capable or worthy than other people. In fact, we often judge ourselves more harshly than we judge others, especially when we make a mistake or feel stressed out. That makes us feel isolated, unhappy, and even more stressed; it may even make us try to feel better about ourselves by denigrating other people.

Rather than harsh self-criticism, a healthier response is to treat yourself with compassion and understanding. According to psychologist Kristin Neff, this "self-compassion" has three main components: mindfulness, a feeling of common humanity, and self-kindness. This exercise walks you through all three of those components when you're going through a stressful experience. Research suggests that people who treat themselves with compassion rather than criticism in difficult times experience greater physical and mental health.

HOW TO DO IT

1. Think of a situation in your life that is difficult and is causing you stress.
2. Call the situation to mind and see if you can physically feel the stress and emotional discomfort in your body.
3. Now, say to yourself, "This is a moment of suffering." This acknowledgment is a form of mindfulness—simply noticing what is going on for you emotionally in the present moment, without judging that experience as good or bad. You can also say to yourself, "This hurts," or, "This is stress." Use whatever statement feels most natural to you.
4. Next, say to yourself, "Suffering is a part of life." This is a recognition of your common humanity with others—that all people have trying experiences, and these experiences give you something in common with the rest of humanity and don't make you abnormal or deficient. Other options for this statement include "Other people feel this way," "I'm not alone," or "We all struggle in our lives."
5. Now, put your hands over your heart, feel the warmth of your hands and

the gentle touch on your chest, and say, "May I be kind to myself." This is a way to express self-kindness. You can also consider whether another phrase would speak to you in that particular situation. Some examples: "May I give myself the compassion that I need," "May I accept myself as I am," "May I learn to accept myself as I am," "May I forgive myself," "May I be strong," and "May I be patient."

This practice can be used any time of day or night. If you practice it in moments of relative calm, it might become easier for you to experience the three parts of self-compassion—mindfulness, common humanity, and self-kindness—when you need them most.

WHY IT WORKS

The three elements in this practice—mindfulness, common humanity, and self-kindness—all play important roles in increasing self-compassion. Mindfulness allows people to step back and recognize that they are experiencing suffering without judging the suffering as something bad that they should avoid. Sometimes, people fail to notice when they are in pain or deny that they are suffering because it brings up feelings of weakness or defeat. Common humanity reminds people of their connection with other people—all of whom suffer at some point in their lives—and eases feelings of loneliness and isolation. Self-kindness is an active expression of caring toward oneself that can help people clarify their intentions for how they want to treat themselves.

Going through these steps in response to a stressful experience can help people replace their self-critical voice with a more compassionate one that comforts and reassures rather than berating them for shortcomings. That makes it easier to work through stress and reach a place of calm, acceptance, and happiness.

WALKING MEDITATION

TIME REQUIRED: 10 MINUTES DAILY FOR AT LEAST A WEEK (EVIDENCE SUGGESTS THAT MINDFULNESS INCREASES THE MORE YOU PRACTICE IT.)

Much of our time is spent rushing from place to place, so preoccupied with our next activity that we don't really notice what we're doing in the present. We risk not really experiencing our life as we live it.

Practicing mindfulness can help. Mindfulness helps us tune into what we're sensing and experiencing in the present moment—it's the ability to pay more careful attention to our thoughts, feelings, and bodily sensations, without judging them as good or bad. Research suggests that it can reduce stress and increase our experience of positive emotions.

One of the basic methods for cultivating mindfulness is a "walking meditation," which involves focusing closely on the physical experience of walking, paying attention to the specific components of each step. With practice, an everyday action that you do automatically, even mindlessly, can become an opportunity for greater focus and awareness—a habit you can try to bring to other mundane activities as well. Some experts recommend alternating the walking meditation with other forms of meditation to keep your practice varied and determine which form feels best for you.

HOW TO DO IT

The steps below are adapted from a guided walking meditation led by mindfulness expert Jon Kabat-Zinn.

1. FINDING A LOCATION: Find a space where you can walk back and forth for 10-15 paces—a place that is relatively peaceful, where you won't be disturbed or even observed (since a slow, formal walking meditation can look strange to people who are unfamiliar with it). You can practice walking meditation either indoors or outside in nature. The space doesn't have to be large since the goal is not to reach a specific destination but rather to practice a very intentional form of walking in which you're mostly retracing your steps.

2. STARTING YOUR STEPS: Walk 10-15 steps in the space you've chosen, and then pause and breathe for as long as you like. When you're ready, turn and walk back in the opposite direction to the other end of the space, where you can pause and breathe again. Then, when you're ready, turn once more and continue with the walk.

3. **The components of each step:** Walking meditation involves very deliberate thinking and execution of a series of actions that you normally do automatically. Breaking these steps down in your mind may feel awkward or even ridiculous. But you should try to notice at least these four basic components of each step:
   - the lifting of one foot
   - the moving of the foot a bit forward of where you're standing
   - the placing of the foot on the ground, heel first
   - the shifting of the weight of the body onto the front leg as the back heel lifts, while the toes of the back foot remain touching the ground, and the cycle continuing as you lift your back foot totally off the ground
   - the back foot swinging forward and lowering, now becoming the front foot
   - the foot making contact with the ground, heel first
   - the weight shifting onto that foot as the body continue moving forward
4. **Speed:** You can walk at any speed, but in Kabat-Zinn's Mindfulness Based Stress Reduction (MBSR) program, walking meditation is slow and involves taking small steps. Most important is that it feels as natural as possible, not exaggerated or stylized.
5. **Hands and arms:** You can clasp your hands behind your back or in front of you, or you can just let them hang at your side—whatever feels most comfortable and natural.
6. **Focusing your attention:** As you walk, try to focus your attention on one or more sensations that you would normally take for granted, such as your breath moving into and out of your body, the movement of your feet and legs or their contact with the ground or floor, your head balanced on your neck and

shoulders, sounds nearby or those caused by the movement of your body, or whatever your eyes take in as they focus on the world in front of you.

7. **WHAT TO DO WHEN YOUR MIND WANDERS:** No matter how much you try to fix your attention on any of these sensations, your mind will inevitably wander. That's okay—it's perfectly natural. When you notice your mind wandering, simply try to re-focus it one of those sensations.

8. **INTEGRATING WALKING MEDITATION INTO YOUR DAILY LIFE:** For many people, slow, formal walking meditation is an acquired taste. The more you practice, even for short periods of time, the more it is likely to grow on you. Keep in mind that you can also bring mindfulness to walking at any speed in your everyday life, and even to running, though of course, the pace of your steps and breath will change. In fact, over time, you can try to bring the same degree of awareness to any everyday activity, experiencing the sense of presence that is available to us at every moment as our lives unfold.

WHY IT WORKS

Walking meditation can help increase our awareness of our internal sensations and our external surroundings, tuning us into experiences that we often miss when we rush on autopilot from place to place. Paying closer attention to the process of walking can also increase our sense of appreciation and enjoyment of our physical bodies. By heightening our awareness of our mental and physical states, walking meditation—like mindfulness in general—can help us gain a greater sense of control over our thoughts, feelings, and actions, allowing us to respond in more constructive ways when we experience negative thoughts or emotions.

REFLECT ON HOW MINDFULNESS LEADS TO GRATITUDE.

CONNECTION

CONNECTION SEEMS TO BE A POWERFUL HAPPINESS BOOSTER: Kids who are more socially connected grow up to be happier—connection even seems to have a stronger impact on happiness than academic achievement.

FEELING CONNECTED MAKES US KIND: In one study, people who thought or wrote about connection were then more likely to be kind and helpful by donating money to charity or planning to volunteer.

CONNECTION IS GOOD FOR OUR HEALTH: Adults with a stronger network of friends live longer, and seniors who are more socially active experience less cognitive decline and disability. In contrast, lonely people tend to have more stress and higher blood pressure.

HUMANS EVOLVED TO BE SOCIAL CREATURES: Our ancestors needed to cooperate and work together to survive. The imprints of our history as a species can be seen in our biology, from the flow of the hormone oxytocin to our keen sense of touch.

CONNECTION CREATES TIES: Social connections come in many different forms, from casual acquaintances to lifelong friends. Having a range of "strong" and "weak" ties makes us healthier and more likely to succeed in life.

Research provided by UC Berkeley's Greater Good Science Center.

REMEMBER CONNECTION

TIME REQUIRED: 10 MINUTES (TRY TO DO THIS PRACTICE AT LEAST ONCE PER WEEK, SELECTING A DIFFERENT EXAMPLE EACH TIME.)

Humans have a strong drive to be kind, but that drive is usually stronger when they feel connected to other people. To help foster that feeling of closeness, this exercise asks you to think about a time when you felt a strong connection to another person and to describe the experience in writing. Research suggests that reflecting on feelings of connection can increase people's motivation to help others, whether by helping a friend or stranger in need, volunteering, or donating money. Helping others can, in turn, increase happiness and improve relationships.

HOW TO DO IT

Try to think of a time when you felt a strong bond with someone in your life. Choose a specific example of an experience you had with this person where you felt especially close and connected to them. This could be a time you had a meaningful conversation, gave or received support, experienced a great loss or success together, or witnessed an historic moment together.

Once you've thought of a specific example, spend a few minutes writing about what happened. In particular, consider the ways in which this experience made you feel close and connected to the other person.

Example 1: ______________________________

____

____

____

____

Example 2: ______________________________

____

____

____

____

Example 3:

Example 4:

Example 5:

WHY IT WORKS

Feeling connected to others is considered to be a fundamental psychological need. When people feel rejected or alone, they may be more likely to focus on themselves and their own unmet needs rather than attending to the needs of others. When people feel connected and cared about, in contrast, they are better able to expend energy helping and caring for others.

By reflecting on times when you've felt a strong connection with others and by striving to cultivate more of these experiences, you are fueling your drive to practice kindness and compassion.

COMPASSIONATE MEDIATION

TIME REQUIRED: 30 MINUTES A DAY FOR TWO WEEKS

Having compassion means that you want others to be free from suffering and you have the urge to help end their suffering. It is not only vital to a kind and just society—research also suggests it is a path to better health and stronger relationships.

Yet cultivating compassion for others—and yourself—can sometimes feel like an emotionally taxing and demanding task. This exercise walks you through a meditation grounded in simple techniques—paying attention to your breath and guided imagery—to help you nurture compassion toward a loved one, yourself, a neutral person, and even an enemy.

HOW TO DO IT

This exercise draws on a guided meditation created by researcher Helen Weng and her colleagues at the Center for Investigating Healthy Minds (CIHM) at the University of Wisconsin, Madison. Dr. Weng called this meditation a "compassion meditation," though a similar kind of meditation is also referred to as a "loving-kindness meditation."

We have included a script of this meditation for you to follow yourself or teach to others. In brackets are the lengths of the pauses in the original guided meditation to give you a sense of how much time to allot to each step.

SETTLING

Settle into a comfortable position and allow yourself to relax.

Take a deep breath and release. [2 SECONDS]

For a few moments, just focus on your breath and clear your mind of worries. Notice when you are breathing in ... and breathing out. Let yourself experience and be aware of the sensations of breathing. [10 SECONDS]

LOVING-KINDNESS & COMPASSION FOR A LOVED ONE

Picture someone who is close to you—someone toward whom you feel a great amount of love. Notice how this love feels in your heart.

Notice the sensations around your heart. Perhaps you feel a sensation of warmth, openness, and tenderness. [10 SECONDS]

Continue breathing, and focus on these feelings as you visualize your loved one. As you breathe out, imagine that you are extending a golden light that holds your warm feelings from the center of your heart. Imagine that the golden light reaches out to your loved one, bringing them peace and happiness. At the same time, silently recite these phrases. [1 MINUTE]

- May you have happiness.
- May you be free from suffering.
- May you experience joy and ease.
- May you have happiness.
- May you be free from suffering.
- May you experience joy and ease.

As you silently repeat these phrases, remember to extend the golden light to your loved one from your heart. Feel with all your heart that you wish your loved one happiness and freedom from suffering.

COMPASSION FOR A LOVED ONE

Now, think of a time when this person was suffering. Maybe they experienced an illness, an injury, or a difficult time in a relationship. [15 SECONDS]

Notice how you feel when you think of their suffering. How does your heart feel? Do the sensations change? Do you continue to feel warmth, openness, and tenderness? Are there other sensations—perhaps an aching sensation? [10 SECONDS]

Continue to visualize your loved one as you breathe. Imagine that you are extending the golden light from your heart to your loved one, and that the golden light is easing their suffering. Extend this light out to them during your exhalation, with the strong heartfelt wish that they be free from their suffering. Recite these phrases to them: [1 MINUTE]

- May you be free from this suffering.
- May you have joy and happiness.
- May you be free from this suffering.
- May you have joy and happiness.

Notice how this feels in your heart. What happened to your heart? Did the sensations change? Did you continue to feel warmth, openness, and tenderness? Were there other sensations—an aching sensation, perhaps? Did you have a wish to take away the other's suffering? [30 SECONDS]

Compassion for Self

Contemplate a time when you suffered yourself. Perhaps you experienced a conflict with someone you care about, did not succeed in something you wanted, or were physically ill. [15 SECONDS]

Notice how you feel when you think of your suffering. How does your heart feel? Do you continue to feel warmth, openness, and tenderness? Are there other sensations—perhaps an aching sensation? [10 SECONDS]

Just as we wish for our loved one's suffering to end, we wish that our own suffering would end. We may also envision our own pain and suffering leaving us so that we may experience happiness.

Continue to visualize yourself as you breathe. Imagine that the golden light emanating from your heart is easing your suffering. With each exhalation, feel the light emanating within you, with the strong heartfelt wish that you be free from your suffering. Silently recite these phrases to yourself. [2 MINUTES]

- May I be free from this suffering.
- May I have joy and happiness.
- May I be free from this suffering.
- May I have joy and happiness.

Again, notice how this feels in your heart. What kind of sensations did you feel? Did they change from when you were envisioning your own suffering? How is this feeling different from when you wished your loved one's suffering to be relieved? Did you feel warmth, openness, and tenderness? Were there other sensations, such as pressure? Did you have a wish to take away your own suffering? [30 SECONDS]

Compassion for a Neutral Person

Now, visualize someone you neither like nor dislike—someone you may see in your everyday life, such as a classmate with whom you are not familiar, a bus driver, or a stranger you pass on the street. [5 SECONDS]

Although you are not familiar with this person, think of how this person may suffer in their own life. This person may also have conflicts with loved ones, struggle with addiction, or suffered illness. Imagine a situation in which this person may have suffered. [30 SECONDS]

Notice your heart center. Does it feel different? Do you feel more warmth, openness, and tenderness? Are there other sensations—perhaps an aching sensation? How does your heart feel different from when you were envisioning your own or a loved one's suffering? [10 SECONDS]

Continue to visualize this person as you breathe. Imagine that you are extending the golden light from your heart to them, and that the golden light is easing their suffering. Extend this light out to them during your exhalation, with the strong heartfelt wish that they be free from suffering. See if this wish can be as strong as the wish for your own or a loved one's suffering to be relieved. Silently recite these phrases to them. [2 MINUTES]

- May you be free from this suffering.
- May you have joy and happiness.
- May you be free from this suffering.
- May you have joy and happiness.

Again, notice how this feels in your heart. Did the sensations change from when you were envisioning this person's suffering? Did you continue to feel warmth, openness, and tenderness? Were there other sensations? Did you have a wish to take away this person's suffering? How were these feelings different from when you were wishing to take away your own or a loved one's suffering? [30 SECONDS]

Compassion for an Enemy

Now, visualize someone with whom you have difficulty in your life. This may be a parent or child with whom you disagree, an ex-partner, a roommate with whom you had an argument, or a coworker with whom you do not get along. [5 SECONDS]

Although you may have negative feelings toward this person, think of how this person has suffered in their own life. This person has also had conflicts with loved ones, dealt with failures, or suffered illness. Think of a situation in which this person may have suffered. [30 SECONDS]

Notice your heart center. Does it feel different? Do you feel more warmth, openness, and tenderness? Are there other sensations— perhaps an aching sensation? How does your heart feel different from when you were envisioning your own or a loved one's suffering? [10 SECONDS]

Continue to visualize this person as you breathe. Imagine that you are extending the golden light from your heart to them, and that the golden light is easing their suffering. Extend this light out to them during your exhalation, with the strong heartfelt wish that they be free from suffering. See if this wish can be as strong as the wish for your own or a loved one's suffering to be relieved. Silently recite these phrases to them. [1 MINUTE]

- May you be free from this suffering.
- May you have joy and happiness.
- May you be free from this suffering.
- May you have joy and happiness.

If you have difficulty in wishing for this person's suffering to be relieved, you may think of a positive interaction you have had with this person that can help you in wishing them joy and happiness. Perhaps there were times when you got along, laughed together, or worked well together. Continue to silently recite these phrases. [2 MINUTES]

- May you be free from this suffering.
- May you have joy and happiness.

Again, notice how this feels in your heart. Did the sensations change? Did you feel warmth, openness, and tenderness? How were these feelings different from when you were wishing for your own or a loved one's suffering to end? Were there other sensations—perhaps a tightness in the chest? Did you have a wish to take away this person's suffering? [30 SECONDS]

Compassion for All Beings

Now that we are almost at the end of this meditation, let's end with a wish for all other beings' suffering to be relieved. Just as you wish to have peace, happiness, and to be free from suffering, so do all beings. [10 SECONDS]

Now, bask in the joy of this open-hearted wish to ease the suffering of all people and how this attempt brings joy, happiness, and compassion in your heart at this very moment.

You have now finished this compassion meditation session.

WHY IT WORKS

This meditation fosters feelings of compassion and concern for others by training people to notice suffering and strive to alleviate it, while at the same time giving people the emotional resources to not feel overwhelmed by the distress caused by that suffering. The researchers who used this compassion meditation in their work argue that the care for others emphasized by the compassion training may have caused participants to see suffering not as a threat to their own well-being but as an opportunity to reap the psychological rewards from achieving an important goal—namely, connecting with someone else and making that person feel better.

By first extending compassion to a loved one and to the self, it becomes easier to extend that same compassion to others, even those you may not like. Extending compassion to people you dislike can help to reduce feelings of hostility and resentment and may lead to improvements in a strained relationship. With practice, this meditation can help bring more peace, joy, and connection to one's own life and to the lives of others.

REFLECT ON HOW CONNECTION LEADS TO GRATITUDE.

CURIOSITY

CURIOSITY HELPS US SURVIVE: The urge to explore and seek novelty helps us understand our constantly changing environment, which may be why our brains evolved to release dopamine and other feel-good chemicals when we encounter new things.

CURIOUS PEOPLE ARE HAPPIER: Studies link curiosity with positive emotions and psychological well-being.

CURIOSITY BOOSTS ACHIEVEMENT: Research indicates that curiosity leads to greater learning, engagement, and performance at school and work.

CURIOSITY CAN EXPAND OUR EMPATHY: When we are curious about others and talk to people outside our usual social circle, we become better able to understand those with lives and worldviews different than our own.

CURIOSITY STRENGTHENS OUR RELATIONSHIPS: In one study, after people asked a stranger personal questions and answered the stranger's own questions, the stranger rated self-identified curious people as more attractive and felt closer to them than less curious people.

CURIOSITY IMPROVES HEALTHCARE: Research has shown that when doctors are genuinely curious about their patients' perspectives, both doctors and patients report less anger and frustration and make better decisions, ultimately increasing the effectiveness of their treatment.

Research provided by UC Berkeley's Greater Good Science Center.

NOTICING NEW THINGS

TIME REQUIRED: VARIES BASED ON LENGTH OF ACTIVITY

Almost every day, we encounter tasks that aren't particularly interesting or enjoyable, but that we have to do anyway. Sometimes we merely go through the motions and take an experience for granted when we could be getting much more out of it. Or we avoid an activity because we think we're not interested or would dislike it, when in fact we are missing out on ways it could enrich our lives.

Research suggests that approaching these situations and others with curiosity can not only make them more enriching but also help us experience more happiness in life. This one very simple curiosity-building exercise can help foster interest and enjoyment in any of these situations.

HOW TO DO IT

Choose an activity that you think you are not interested in or even dislike, such as a daily task or chore. Alternatively, try something out of the ordinary for you, like listening to a genre of music that's not your favorite, eating a food you disliked as a child, watching a sport you think is boring, or trying a hobby you've never found intriguing.

Try to let go of any expectations, positive or negative, that you have about the experience. Simply keep an open, curious mind.

While engaged in the activity, take note of at least three new things about it that you have never noticed before.

You may find your preconceived notions changing, opening up new possibilities for interest and enjoyment in your life. Even if not, you will have added a few new and interesting things to your catalog of experience.

ACTIVITY 1: ______________________________

ACTIVITY 2: ______________________________

ACTIVITY 3: ______________________________

ACTIVITY 4: ______________________________

ACTIVITY 5: ______________________________

WHY IT WORKS

Sometimes routines or preconceived notions cause us to lose interest in an activity or experience, often before we've ever really given it a chance. By forcing ourselves to pay closer attention to it, we may cease to take its strengths for granted and start to appreciate its value. In her book *On Becoming an Artist*, psychologist Ellen Langer writes that, as she ran studies using this exercise, she found that "it became clear that taking notice of things expands our appreciation of them."

REFLECT ON HOW CURIOSITY LEADS TO GRATITUDE.

FORGIVENESS

FORGIVENESS IMPROVES OUR HEALTH: When we hold onto grudges, our blood pressure and heart rate spike; when we forgive, our stress levels drop.

FORGIVENESS MAKES US HAPPIER: Research suggests that if you forgive someone today, you'll feel happier tomorrow.

FORGIVENESS IS GOOD FOR RELATIONSHIPS (MOST OF THE TIME): Spouses who are more forgiving and less vindictive are better at resolving conflicts and have stronger, more satisfying relationships. However, when more forgiving spouses are frequently mistreated by their partner, they become less satisfied with their marriage.

FORGIVENESS IS ASSOCIATED WITH KINDNESS: People who are forgiving are more likely to want to volunteer and donate money to charity, and they feel more connected to other people in general.

FORGIVENESS CAN HELP HEAL THE WOUNDS OF WAR: A research-based forgiveness training program in Rwanda, for instance, was linked to reduced trauma and more positive attitudes between the Hutus and Tutsis there.

Research provided by UC Berkeley's Greater Good Science Center.

STEPS TO FORGIVENESS

TIME REQUIRED: EACH PERSON WILL FORGIVE AT THEIR OWN PACE; MOVE THROUGH THE STEPS BELOW BASED ON WHAT WORKS FOR YOU

We have all suffered hurt and betrayal. Choosing to forgive is a way to release the distress that arises again and again from the memory of these incidents—but forgiveness is often a long and difficult process.

This exercise outlines several steps that are essential to the process of forgiveness, breaking it down into manageable components. These steps were created by Robert Enright, PhD, one of the world's leading forgiveness researchers. Although the exact process of forgiveness may look different for different people, anyone can draw upon Dr. Enright's basic principles. In certain cases, it may help to consult a trained clinician, especially if you are working through a traumatic event.

HOW TO DO IT

1. Make a list of people who have hurt you deeply enough to warrant the effort to forgive. You can do this by asking yourself, on a 1-to-10 scale, "How much pain do I have regarding the way this person treated me?" with 1 involving the least pain (but still significant enough to justify the exercise) and 10 involving the most pain. Then, order the people on this list from least to most painful. Start with the person lowest on this hierarchy (least painful).
2. Consider one offense by the first person on your list. Ask yourself, "How has this person's offense negatively impacted by life?" Reflect on the psychological and physical harm it may have caused. Consider how your view of humanity and trust of others may have changed as a result of this offense. Recognize that what happened was not okay, and allow yourself to feel any negative emotions that come up.
3. When you're ready, make a decision to forgive. Deciding to forgive involves coming to terms with what you will be doing as you forgive—extending an act of mercy toward the person who has hurt you. When we offer this mercy, we deliberately try to reduce resentment (persistent ill will) toward this person and instead offer them kindness,

respect, generosity, or even love. It is important to emphasize that forgiveness does not involve excusing the person's actions, forgetting what happened, or tossing justice aside. Justice and forgiveness can be practiced together. Another important caveat: To forgive is not the same as to reconcile. Reconciliation is a negotiation strategy in which two or more people come together again in mutual trust. You may choose not to reconcile with the person you are forgiving.

4. Start with cognitive exercises. Ask yourself these questions about the person who has hurt you: What was life like for this person while growing up? What wounds did they suffer from others that could have made them more likely to hurt you? What kinds of extra pressures or stresses were in this person's life at the time they offended you? These questions are not meant to excuse or condone, but rather to better understand the other person's areas of pain, those areas that make them vulnerable and human. Understanding why people commit destructive acts can also help us find more effective ways of preventing further destructive acts from occurring in the future.
5. Be aware of any little movement of your heart through which you begin to feel even slight compassion for the person who offended you. This person may have been confused, mistaken, and misguided. They may deeply regret their actions. As you think about this person, notice if you start to feel softer emotions toward them.
6. Try to consciously bear the pain that they caused you so that you do not end up throwing that pain back onto the one who offended you, or even toward unsuspecting others, such as loved ones who were not the ones who wounded you in the first place. When we are emotionally wounded, we tend to displace our pain onto others. Please be aware of this so that you are not perpetuating a legacy of anger and injuries.

7. Think of a gift of some kind that you can offer to the person you are trying to forgive. Forgiveness is an act of mercy—you are extending mercy toward someone who may not have been merciful toward you. This could be through a smile, a returned phone call, or a good word about them to others. Always consider your own safety first when extending kindness and good will toward this person. If interacting with this person could put you in danger, find another way to express your feelings, such as by writing in a journal or engaging in a practice such as compassion meditation.
8. Finally, try to find meaning and purpose in what you have experienced. For example, as people suffer from the injustices of others, they often realize that they themselves become more sensitive to others' pain. This, in turn, can give them a sense of purpose toward helping those who are hurting. It may also motivate them to work toward preventing future injustices of a similar kind.

Once you complete the forgiveness process with one person on your list, select the next person and continue up the list until you are forgiving the person who hurt you the most.

WHY IT WORKS

Forgiveness is a long and often challenging process. These steps may help along the way by providing concrete guidelines. Specifically, they may help you narrow and understand whom to forgive—to name and describe your pain, to understand the difference between forgiving and excusing or reconciling. By thinking about the person who has caused you pain in a novel way, you may begin to feel some compassion for them, facilitating forgiveness and reducing the ill will you hold toward this person. These steps also attune you to residual pain from your experience and encourage you to find meaning and some positivity in it.

REFLECT ON HOW FORGIVENESS LEADS TO GRATITUDE.

WONDER

WONDER BROADENS OUR HORIZONS: Psychologists define "awe" as the feeling we get when we come across something so strikingly vast in number, scope, or complexity that it alters the way we understand the world.

WONDER IS ASSOCIATED WITH BETTER HEALTH: Based on their physiological makeup, people who experience feelings of awe and wonder seem to be at lower risk for depression, heart disease, diabetes, and Alzheimer's disease.

WONDER MAKES US FEEL LESS RUSHED: When we're in the presence of something tremendous—like a whale or a waterfall—we are less impatient and feel like we have more time, which makes us more willing to volunteer our time to help others.

WONDER MAKES US HAPPIER: When people experience a sense of wonder, they report feeling more satisfied with their lives.

WONDER MAKES US FEEL LIKE WE'RE PART OF SOMETHING BIGGER THAN OURSELVES: When we're in the presence of something wonder-inspiring, we put less emphasis on our ego and feel connected to a greater whole.

WONDER INCREASES KINDNESS: After spending time looking up at an impressive eucalyptus grove, study participants were more helpful to someone in need and felt less entitled to a reward.

Research provided by UC Berkeley's Greater Good Science Center.

WONDER NARRATIVE

TIME REQUIRED: 15 MINUTES

This exercise asks you to recall and describe a time when you experienced awe. Awe is an emotion that is elicited by experiences that challenge and expand our typical way of seeing the world—it provides the kind of change in perspective that can elicit creative solutions to challenging problems. Research suggests that awe involves sensing the presence of something greater than oneself, along with decreased self-consciousness and a decreased focus on minor, everyday concerns. Experiences of awe have been shown to expand people's perception of time and improve life satisfaction.

HOW TO DO IT

Think back to a time when you felt a sense of awe regarding something you witnessed or experienced. Awe has been defined as a response to things that are perceived as vast and overwhelming and that alter the way we understand the world. This sense of vastness can be physical (e.g. a panoramic view from a mountaintop) or psychological (e.g. a brilliant idea). People may experience awe when they are in the presence of a beautiful natural landscape or work of art, when they watch a moving speech or performance, when they witness an act of great altruism, or when they have a spiritual or religious experience.

Try to think of the most recent experience you've had that involved the feeling of awe. Once you identify something, describe it here in as much detail as possible.

WHY IT WORKS

Taking time to reflect on past experiences of awe can help people break up their routines and challenge themselves to think in new ways. Evoking feelings of awe may be especially helpful when people are feeling bogged down by day-to-day concerns. Research suggests that awe has a way of lifting people outside of their day-to-day selves and connecting them with something larger and more significant. This sense of broader connectedness and purpose can help relieve negative moods and improve happiness.

WONDER STORY

TIME REQUIRED: 10 MINUTES TO READ THIS STORY (FOR A REGULAR DOSE OF AWE, MAKE TIME TO READ A STORY LIKE THIS AT LEAST ONCE PER WEEK.)

It's easy to feel bogged down by daily routines and mundane concerns, stifling our sense of creativity and wonder. Feeling awe can reawaken those feelings of inspiration.

Awe is induced by experiences that challenge and expand our typical way of seeing the world, often because we sense that we're in the presence of something greater than ourselves. Research suggests that experiencing awe improves people's satisfaction with life, makes them feel like they have more time, makes them feel less self-conscious, and reduces their focus on trivial concerns.

But in our everyday lives, we might not regularly encounter things that fill us with awe. That's where this practice comes in. It's a way to experience awe even if you can't make it to an inspiring vista or museum. It involves reading a story that has been shown to induce awe, giving you the chance to infuse even your most mundane days with a dose of wonder.

HOW TO DO IT

Set aside at least 10 minutes to read the story below.

Of course, reading a story like this is not the only way to elicit awe, and there are many different types of stories that could have this effect. The stories and other stimuli that inspire awe tend to share two key features:

1. They involve a sense of vastness that puts into perspective your own relatively small place in the world. This vastness could be either physical (e.g., a panoramic view from a mountaintop) or psychological (e.g., an exceptionally courageous or heroic act of conscience).
2. They alter the way you understand the world. For instance, they might make your everyday concerns seem less important, or they might expand your beliefs about the reaches of human potential.

READ THE STORY BELOW TO EXPERIENCE THESE DIMENSIONS OF AWE.
Imagine you're getting ready to go on a trip to Europe. Although you've seen parts of Europe in photos and on television, you know that seeing things in person will be a completely different experience. You're particularly excited to begin the trip in one of the most inspiring capitals of the continent—the magnificent city of Paris.

As soon as you arrive in Paris, you're overwhelmed by the grandeur and beauty of the historic city. The sights, smells, and sounds are like nothing you have ever experienced. Everywhere you look there is something new to capture your imagination. Scanning the view from left to right, you're surrounded by beautiful buildings. Famous museums and churches beckon for you to absorb the stories of their rich past, while centuries-old hotels and city buildings exude majesty and history. As you pass by them, you're amazed by the elaborate architectural designs and the ornate details. Between two of the buildings, you catch a glimpse of the Eiffel tower in the distance. Seeing it for the first time in person, your eyes widen and your senses feel wide awake. Although it looks small from where you are, the incredible height of the tower becomes clear as you walk toward it.

Standing a block from the tower, you're overwhelmed by the sheer size and grandeur of the structure. The intricately woven beams of steel rise high from the ground, and you feel completely dwarfed standing next to it. You look up, but you can't even see the top. The magnitude of the tower is enormous and it feels even more amazing being there in person than you could have ever imagined. The metal beams rising from the ground are larger than the biggest tree trunks you've ever seen. You touch them: As your hands come in contact with the cold metal, you feel the presence of something greater than yourself, not just physically, but in human history. You can't believe that something so tremendous was built by man.

You take the elevator to the top. During the ride you can't help but think back to the first time you saw the Grand Canyon—that moment when everything around you stops as you try to comprehend what's in front of you. Finally, the elevator doors begin to open, and there it is—Paris all around you. As you take in the overwhelming sight, your mouth opens and you catch your breath. The famed City of Lights stretches for miles in all directions around you, yet from this vantage point the hustle and bustle below cease to exist. As your body is enveloped by a strong feeling of wonder, you scan the enormous panorama and try to take in everything that's in front of you. You lose yourself in the beauty of the sight.

JOURNAL ON A TIME YOU FELT AWE.

WHY IT WORKS

Taking time out to experience awe can help people break up their routine and challenge themselves to think in new ways. Evoking feelings of awe may be especially helpful when people are feeling bogged down by day-to-day concerns. Research suggests that awe has a way of lifting people outside of their narrower senses of self, connecting them with something larger and more significant. This sense of broader connectedness and purpose can help relieve negative moods and improve happiness.

REFLECT ON HOW WONDER LEADS TO GRATITUDE.

REFERENCES

CREATIVITY

Feist, Gregory. "A Meta-Analysis of Personality in Scientific and Artistic Creativity." *Personality and Social Psychology Review* 2, no. 4 (November 1998): 290-309.

U.S. Department of Justice Office of Juvenile Justice and Delinquency Prevention. *The YouthARTS Development Project*, by Heather J. Clawson and Kathleen Coolbaugh. Washington, D.C.: Juvenile Justice Bulletin, 2001.

Favara-Scacco, Cinzia, Giuseppina Smirne, Gino Schiliro, and Andrea Di Cataldo. "Art Therapy as Support for Children with Leukemia During Painful Procedures." *Medical and Pediatric Oncology* 36, no. 4 (April 2001): 474-480.

People & Stories. https://peopleandstories.org/

Petrie, Keith, Iris Fontanilla, Mark Thomas, Roger Booth, and James Pennebaker. "Effect of Written Emotional Expression on Immune Function in Patients With Human Immunodeficiency Virus Infection: A Randomized Trial." *Psychosomatic Medicine* 66, no. 2 (April 2004): 272-275.

Hickson, Joyce and Warren Housley. "Creativity in Later Life." *Educational Gerontology* 23, no. 6 (1997): 539-547.

Livingston, Judith A., PhD. "Something Old and Something New: Love, Creativity, and the Enduring Relationship." *Bulletin of the Menninger Clinic* 63, no. 1 (Winter 1999): 40-52.

MEANINGFUL PHOTOS

Steger, Michael, Yerin Shim, Jennifer Barenz, and Joo Yeon Shin. "Through the Windows of the Soul: A Pilot Study Using Photography to Enhance Meaning in Life." *Journal of Contextual Behavioral Science* 3, no. 1 (January 2014): 27-30.

College students were instructed to take 9-12 photographs of things that they felt made their life meaningful; one week later, they viewed and wrote about each photograph. They completed a battery of questionnaires before and after this exercise. Afterward, they reported feeling like they had more meaning in their lives, greater life satisfaction, and more positive emotion than they had beforehand.

Michael Steger, PhD, Colorado State University

PURPOSE

Zhang, Neil Si-Jia. "Can Purpose Keep You Alive?" *Greater Good Magazine*, August 2014.

Suttie, Jill. "A Healthier Kind of Happiness." *Greater Good Magazine*, September 2013.

Suttie, Jill and Jason Marsh. "Is a Happy Life Different from a Meaningful One?" *Greater Good Magazine*, February 2014.

GOAL VISUALIZATION

Sergeant, Susan and Myriam Mongrain. "An Online Optimism Intervention Reduces Depression in Pessimistic Individuals." *Journal of Consulting and Clinical Psychology* 82, no. 2 (April 2014): 263-274.

Participants who completed this Goal Visualization exercise (along with the Silver Linings practice) daily for three weeks reported greater engagement in life and less dysfunctional thinking (e.g., believing that small failures make you a failure as a person) at the end of the study than they had at the start of it. Participants who had a tendency to be pessimistic especially benefited from the exercises and showed fewer depressive symptoms afterward. However, these effects seemed to wear off two months later, suggesting the need to repeat this practice periodically.

Myriam Mongrain, PhD, York University, United Kingdom

BEST POSSIBLE SELF

Sheldon, Kennon and Sonja Lyubomirsky. "How To Increase and Sustain Positive Emotion: The Effects of Expressing Gratitude and Visualizing Best Possible Selves." *Journal of*

Positive Psychology 1, no. 2 (2006): 73-82.
People who completed the Best Possible Self exercise daily for two weeks showed increases in positive emotions right after the two-week study ended. Those who kept up with the exercise even after the study was over continued to show increases in positive mood one month later.

Laura A. King, PhD, University of Missouri
Jeffrey Huffman, M.D., Harvard Medical School, Massachusetts General Hospital

GENEROSITY

Remembering Connection

Pavey, Louisa, Tobias Greitemeyer, and Paul Sparks. "Highlighting Relatedness Promotes Prosocial Motives and Behavior." *Personality and Social Psychology Bulletin* 37, no. 7 (April 2011): 905-917.
Some study participants reflected on a time when they felt a strong bond with someone else; other participants wrote about a time when they felt especially competent or autonomous. Compared with those in the other groups, the participants who reflected on their experience of closeness reported greater feelings of connectedness and concern for others. What's more, they also reported a stronger intention to carry out a variety of altruistic behaviors over the next six weeks, including giving money to charity and going out of their way to help a stranger in need.

Find Commonalities

Levine, Mark, Amy Prosser, David Evans, and Stephen Reicher. "Identity and Emergency Intervention: How Social Group Membership and Inclusiveness of Group Boundaries Shape Helping Behavior." *Personality and Social Psychology Bulletin* 31, no. 4 (April 2005): 443-453.
Participants were more likely to help a fallen jogger when the jogger was a fellow fan of the same soccer team than when the jogger was a fan of a rival team (as indicated by their shirt). But when participants were reminded of a shared identity with the fallen rival (being a soccer fan), they were more likely to help than they were to help a non-fan.

Leary, Mark, Jessica Tipsord, and Eleanor Tate. "Allo-Inclusive Identity: Incorporating the Social and Natural Worlds Into One's Sense of Self." In *Transcending Self-Interest: Psychological Explorations of the Quiet Ego*, edited by Heidi Wayment and Jack Bauer, 137-147. Washington: American Psychological Association, 2008).
Participants who reported feeling a greater sense of connection to other people, regardless of group distinctions, and to the natural world at large also reported less egocentricity, more concern for others, and less interest in having power over others.

Random Acts of Kindness

Lyubomirsky, Sonja, Kennon Sheldon, and David Schkade. "Pursuing Happiness: The Architecture of Sustainable Change." *Review of General Psychology* 9, no. 2 (June 2005): 111-131.
Study participants who performed five acts of kindness every week for six weeks saw a significant boost in happiness, but only if they performed their five acts in a single day rather than spread out over each week. This may be because many acts of kindness are small, so spreading them out might make them harder to remember and savor.

Sonja Lyubomirsky, PhD, University of California, Riverside

ENERGY

Suttie, Jill. "What Are Your Strengths?" *Greater Good Magazine*, July 2015.

Park, Nansook, Christopher Peterson, and Martin E. P. Seligman. "Strengths of Character and Well-Being." *Journal of Social and Clinical Psychology* 23, no. 5 (2004): 603-619.

Peterson, Christopher, Willibald Ruch, Ursula Beermann, Nansook Park, and Martin E. P. Seligman. "Strengths of Character, Orientations to Happiness, and Life Satisfaction." *Journal of Positive Psychology* 2, no. 3 (July 2007): 149-156.

Peterson, Christopher, Nansook Park, Nicholas Hall, and Martin E. P. Seligman. "Zest and

Work." *Journal of Organizational Behavior* 30, no. 2 (January 2009): 161-172.

Carter, Christine. "The Benefits of Optimism." *Greater Good Magazine*, April 2008.

Wonder Video

Rudd, Melanie, Kathleen Vohs, and Jennifer Aaker. "Awe Expands People's Perception of Time, Alters Decision Making, and Enhances Well-Being." *Psychological Science* 23, no. 10 (August 2012): 1130-1136.

In three experiments, participants were induced to feel awe—such as by watching an awe-inspiring video—as well as other emotions. People who experienced awe felt that they had more time available to themselves, were less impatient, were more willing to volunteer their time to help others, preferred having positive experiences over material products, and reported greater life satisfaction.

Melanie Rudd, PhD, University of Houston

FOCUS

Killingsworth, Matt. "Does Mind-Wandering Make You Unhappy?" *Greater Good Magazine*, July 2013.

Frederickson, Barbara. "Are You Getting Enough Positivity in Your Diet?" *Greater Good Magazine*, June 2011.

Hasenkamp, Wendy. "How to Focus a Wandering Mind." *Greater Good Magazine*, July 2013.

Moore, Adam, Thomas Gruber, Jennifer Derose, and Peter Malinowski. "Regular, Brief Mindfulness Meditation Practice Improves Electrophysiological Markers of Attentional Control." *Frontiers in Human Neuroscience* 6, no. 18 (February 2012).

Marsh, Jason and Bernie Wong. "How Meditation is Good for Mind and Body." *Greater Good Magazine*, June 2011.

Tangney, June, Roy Baumeister, and Angie Luzio Boone. "High Self-Control Predicts Good Adjustment, Less Pathology, Better Grades, and Interpersonal Success." *Journal of Personality* 72, no. 2 (October 2008): 271-324.

Sheaffer, Beverly, Jeannie Golden, and Paige Averett. "Facial Expression Recognition Deficits and Faulty Learning: Implications for Theoretical Models and Clinical Applications." *International Journal of Behavioral Consultation and Therapy* 5, no. 1 (March 2009): 31-55.

Raisin Meditation

Praissman, Sharon. "Mindfulness-based Stress Reduction: A Literature Review and Clinician's Guide." *Journal of the American Academy of Nurse Practitioners* 20, no. 4 (April 2008): 212-216.

A review of research published between 2000 and 2006 concluded that the Mindfulness-Based Stress Reduction Program (MBSR), an eight-week training program that includes the raisin meditation described above, developed by Jon Kabat-Zinn at the University of Massachusetts Medical School, is an effective treatment for reducing the stress and anxiety that accompanies daily life and chronic illness.

"Eating One Raisin: A First Taste of Mindfulness." Extension Service, West Virginia University. Adapted from: Williams, M., Teasdale, J., Zindel, S., & Kabat-Zinn, J. (2007).

Williams, Mark, John Teasdale, Zindel Segal, and Jon Kabat-Zinn. *The Mindful Way through Depression: Freeing Yourself from Chronic Unhappiness*. New York: Guilford Press, 2007.

Body Scan

Carmody, James and Ruth Baer. "Relationships Between Mindfulness Practice and Levels of Mindfulness, Medical and Psychological Symptoms, and Well-being in a Mindfulness-based Stress Reduction Program." *Journal of Behavioral Medicine* 31, no. 1 (February 2008): 23-33.

Participants who attended eight weekly sessions of the Mindfulness-Based Stress Reduction (MSBR) program showed increases in mindfulness and well-being at the end of the eight weeks, and decreases in stress and symptoms of mental illness. Time spent engaging in the body scan in particular was associated with greater levels of two components of mindfulness—observing thoughts, feelings, and physical sensations, and non-reacting to stress—and with increased psychological well-being.

Diana Winston, PhD, UCLA Mindful Awareness Research Center

Steven D. Hickman, PsyD, UC San Diego Center for Mindfulness

COURAGE

Pury, Cynthia, Robin Kowalski, and Jana Spearman. "Distinctions Between General and Personal Courage." *Journal of Positive Psychology* 2, no. 2 (April 2007): 99-114.

Muris, Peter, Birgit Mayer, and Tinke Schubert. "'You Might Belong in Gryffindor': Children's Courage and Its Relationships to Anxiety Symptoms, Big Five Personality Traits, and Sex Roles." *Child Psychiatry and Human Development* 41, no. 2 (October 2009): 204-213.

Overcoming Fear

Schiller, Daniela, Marie-H Monfils, Candace Raio, David Johnson, Joseph LeDoux, and Elizabeth Phelps. "Preventing the Return of Fear in Humans Using Reconsolidation Update Mechanisms." *Nature* 463 (January 2010): 49-53.

People received a mild shock each time they saw a blue square on a computer screen, conditioning them to fear the blue square; evidence of their increased fear came from a subtle measure of increased sweat on their skin. The next day, they underwent "extinction training"—that is, they were repeatedly exposed to the blue square again, but this time without the shocks. After that, they showed a significant decrease in their fear response to the blue square, an effect that persisted a year later.

GRATITUDE

John-Henderson, Neha and Janelle Caponigro. "Team Sports Boost Mental Health." *Greater Good Magazine*, December 2010.

Marsh, Jason. "Tips for Keeping a Gratitude Journal." *Greater Good Magazine*, November 2011.

Gratitude Letter

Seligman, Martin, Tracy Steen, Nansook Park, and Christopher Peterson. "Positive Psychology Progress: Empirical Validation of Interventions." *American Psychologist* 60, no. 5 (August 2005): 410-421.

When researchers tested five different exercises, the gratitude visit showed the greatest positive effect on participants' happiness one month later; however, six months after the visit, their happiness had dropped back down to where it was before. This is why some researchers suggest doing this exercise once every six weeks or so. Also, 2009 research led by Jeffrey Froh found that adolescents who don't generally experience positive emotions showed a significant boost in positive emotions two months after doing a gratitude visit. Research suggests that while there are benefits simply to writing the letter, you reap significantly greater benefits from delivering and reading it in person.

Sonja Lyubomirsky, PhD, University of California, Riverside

Kristin Layous, PhD, Stanford University

Martin Seligman, PhD, University of Pennsylvania

Gratitude Journal

Emmons, Robert and Michael McCullough. "Counting Blessings Versus Burdens: An Experimental Investigation of Gratitude and Subjective Well-being in Daily Life." *Journal of Personality and Social Psychology* 84, no. 2 (February 2003): 377-389.

Participants who kept a gratitude journal weekly for 10 weeks or daily for two weeks experienced more gratitude, positive moods, optimism about the future, and better sleep.

Robert Emmons, PhD, University of California, Davis

Sonja Lyubomirsky, PhD, University of California, Riverside

HAPPINESS

Greater Good Magazine. "What Is Happiness?" Greater Good Science Center, University of California, Berkeley. https://greatergood.

berkeley.edu
Gruber, June. "Four Ways Happiness Can Hurt You." *Greater Good Magazine*, May 2012.
Stutzer, Alois and Bruno Frey. "Does Marriage Make People Happy, Or Do Happy People Get Married?" *Journal of Socio-Economics* 35, no. 2 (April 2006): 326-347.
Lyubomirsky, Sonja, Laura King, and Ed Diener. "The Benefits of Frequent Positive Affect: Does Happiness Lead to Success?" *Psychological Bulletin* 131, no. 6 (November 2005): 803-855.
Weissman, Jordan. "This Study on Happiness Convinced a CEO to Pay All of His Employees At Least $70,000 a Year." *Slate*, April 2015.

Three Good Things

Seligman, Martin, Tracy Steen, Nansook Park, and Christopher Peterson. "Positive Psychology Progress: Empirical Validation of Interventions." *American Psychologist* 60, no. 5 (August 2005): 410-421.

Visitors to a website received instructions for performing this exercise. Writing about three good things was associated with increased happiness immediately afterward, as well as one week, one month, three months, and six months later.

Jeffrey Huffman, M.D., Harvard Medical School, Massachusetts General Hospital
Sonja Lyubomirsky, PhD, University of California, Riverside

Positive Events

Peterson, Christopher, Nansook Park, and Martin Seligman. "Orientations to Happiness and Life Satisfaction: The Full Life Versus the Empty Life." *Journal of Happiness Studies* 6, no. 1 (March 2005): 25-41.

Seeking happiness through pleasurable, engaging, and meaningful activities was found to predict life satisfaction in a sample of 845 adults. Activities that involved deep mental engagement or meaningful pursuits were more strongly associated with happiness than pleasure-seeking activities, but the combination of all three types of activity was associated with the highest levels of life satisfaction.

Huffman, Jeff, Christina DuBois, Brian Healy, Julia Boehm, Todd Kashdan, Christopher Celano, John Denninger, and Sonja Lyubomirksy. "Feasibility and Utility of Positive Psychology Exercises for Suicidal Inpatients." *General Hospital Psychiatry* 36, no. 1 (January 2014): 88-94.

Psychiatric patients hospitalized for suicidal thoughts or behaviors reported increased optimism and decreased hopelessness after completing this exercise.

Jeffrey Huffman, M.D., Harvard Medical School, Massachusetts General Hospital

PATIENCE

Kabat-Zinn, Jon. "What Is Mindfulness?" *Greater Good Magazine*, May 2010.
Spence, Janet, Robert Helmreich, and Robert Pred. "Impatience versus achievement strivings in the Type A pattern: Differential effects on students' health and academic achievement." *Journal of Applied Psychology* 72, no. 4 (December 1987): 522-528.
Wheeler, Sarah. "Can Mindfulness Help Kids Learn Self-Control?" *Greater Good Magazine*, April 2014.
DeSteno, David, Ye Li, Leah Dickens, and Jennifer Lerner. "Gratitude: A Tool for Reducing Economic Impatience." *Psychological Science* 25, no. 6 (April 2014): 1262-1267.

Mindful Breathing

Arch, Joanna and Michelle Craske. "Mechanisms of Mindfulness: Emotion Regulation Following a Focused Breathing Induction." *Behaviour Research and Therapy* 44, no. 12 (2006): 1849-1858.

Participants who completed a 15-minute focused breathing exercise (similar to the mindful breathing exercise described above) reported less negative emotion in response to a series of slides that displayed negative images, compared with people who didn't complete the exercise. These results suggest that the focused breathing exercise helps to improve participants' ability to regulate their emotions.

Diana Winston, PhD, UCLA Mindful Awareness Research Center

Active Listening

Weger, Harry, Gina Castle Bell, Elizabeth Minei, and Melissa Robinson. "The Relative Effectiveness of Active Listening in Initial Interactions." *International Journal of Listening* 28, no. 1 (January 2014): 13-31.

Participants had brief conversations (about their biggest disappointment with their university) with someone trained to engage in active listening, someone who gave them advice, or someone who gave simple acknowledgments of their point of view. Participants who received active listening reported feeling more understood at the end of the conversation.

Instructions adapted from: Markman, H., Stanley, S., & Blumberg, S.L. (1994). *Fighting for Your Marriage*. San Francisco: Josey-Bass Publishers.

LOVE

Luerssen, Anna. "Can We Really Make Love Last?" *Greater Good Magazine*, August 2009.

Graham, Linda. "Love Story: A Review of The Chemistry of Connection." *Greater Good Magazine*, July 2009.

Saxton, Kat, Jason Marsh, Aaron Shaw, and Erica Lee. "Why It Helps to Think Your Partner's Hot." *Greater Good Magazine*, May 2010.

Mental Subtraction

Koo, Minkyung, Sara Algoe, Timothy Wilson, and Daniel Gilbert. "It's a Wonderful Life: Mentally Subtracting Positive Events Improves People's Affective States, Contrary to Their Affective Forecasts." *Journal of Personality and Social Psychology* 95, no. 5 (November 2008): 1217-1224.

Participants were asked to think about what their lives would have been like if a positive event, such as meeting a romantic partner, had never happened; other participants either simply thought about the event or thought about how it was not surprising that the event had happened. The participants who practiced "mental subtraction"—they considered their lives without the positive event— reported feeling more positive states and more gratitude than the other participants did.

Avoiding the Four Horsemen

Over, Harriet and Malinda Carpenter. "Eighteen-Month-Old Infants Show Increased Helping Following Priming with Affiliation." *Psychological Science* 20, no. 10 (September 2009): 1189-1193.

Eighteen-month old children saw a series of photos that had different household objects in the foreground; for some of these children, in the background were two small dolls facing each other—a subtle reminder of connection. But for other children, in the background were two stacks of blocks, a single doll standing alone, or two dolls turned away from each other. After viewing the photos, all of the children had the opportunity to help an adult in need. The children who had seen the subtle reminder of connection were three times more likely to help the adult.

Pavey, Louisa, Tobias Greitemeyer, and Paul Sparks. "Highlighting Relatedness Promotes Prosocial Motives and Behavior." *Personality and Social Psychology Bulletin* 37, no. 7 (April 2011): 905-917.

People who read words associated with human connectedness were more interested in volunteering for a charity and were more likely to donate money to a charity.

MINDFULNESS

Greater Good Magazine. "What Is Mindfulness?" Greater Good Science Center, University of California, Berkeley. https://greatergood.berkeley.edu

Davidson, Richard, Jon Kabat-Zinn, Jessica Schumacher, Melissa Rosenkranz, Daniel Muller, Saki Santorelli, Ferris Urbanowaki et al. "Alterations in Brain and Immune Function Produced by Mindfulness

Meditation." *Psychosomatic Medicine* 65, no. 4 (July 2003): 564-570.

Keng, Shian-Ling, Moria Smoski, and Clive Robins. "Effects of Mindfulness on Psychological Health: A Review of Empirical Studies." *Clinical Psychology Review* 31, no. 6 (August 2011): 1041-1056.

Weinstein, Netta, Kirk Brown, and Richard Ryan. "A Multi-Method Examination of the Effects of Mindfulness on Stress Attribution, Coping, and Emotional Well-Being." *Journal of Research in Personality* 43, no. 3 (June 2009): 374-385.

John-Henderson, Neha. "Is Mindfulness as Good as Antidepressants?" *Greater Good Magazine*, March 2011.

Marsh, Jason. "A Little Meditation Goes a Long Way." *Greater Good Magazine*, February 2011.

Simon-Thomas, Emiliana. "Meditation Makes Us Act with Compassion." *Greater Good Magazine*, April 2013.

Marsh, Jason. "How to Train the Compassionate Brain." *Greater Good Magazine*, May 2013.

Marsh, Jason and Bernie Wong. "How Meditation is Good for Mind and Body." *Greater Good Magazine*, June 2011.

Zeidan, Fadel, Susan Johnson, Bruce Diamond, Zhanna David, and Paula Goolkasian. "Mindfulness Meditation Improves Cognition: Evidence of Brief Mental Training." *Consciousness and Cognition* 19, no. 2 (April 2010): 597-605.

Moore, Adam, Thomas Gruber, Jennifer Derose, and Peter Malinowski. "Regular, Brief Mindfulness Meditation Practice Improves Electrophysiological Markers of Attentional Control." *Frontiers in Human Neuroscience* 6, no. 18 (February 2012).

Campbell, Emily. "Mindfulness in Education Research Highlights." *Greater Good Magazine*, September 2014.

Flook, Lisa, Simon Goldberg, Laura Pinger, Katherine Bonus, and Richard Davidson. "Mindfulness for Teachers: A Pilot Study to Assess Effects on Stress, Burnout, and Teaching Efficacy." *Mind, Brain, and Education* 7, no. 3 (August 2013): 182-195.

Self-Compassion Break

Neff, Kristin and Christopher Germer. "A Pilot Study and Randomized Controlled Trial of the Mindful Self-Compassion Program." *Journal of Clinical Psychology* 69, no. 1 (January 2013): 28-44.

Participants in an eight-week Mindful Self-Compassion (MSC) program, which included practicing the self-compassion break, among other exercises, reported feeling greater self-compassion at the end of the program than they had at the beginning. Their self-compassion at the end of the eight weeks was also greater than that of a comparison group that didn't participate in the program. The MSC participants also reported greater mindfulness and happiness, and lower depression, anxiety, and stress.

Kristin Neff, PhD, University of Texas, Austin
Center for Mindful Self-Compassion

Walking Meditation

Grossman, Paul, Ludger Niemann, Stefan Schmidt, and Harald Walach. "Mindfulness-Based Stress Reduction and Health Benefits, A Meta-Analysis." *Journal of Psychosomatic Research* 57, no. 1 (July 2004): 35-43.

A meta-analysis of 20 published studies concluded that the Mindfulness-Based Stress Reduction Program (MBSR), an eight-week training program that includes the walking meditation described above, is effective in improving physical symptoms and psychological well-being among individuals experiencing physical and mental illness (e.g., cancer, heart disease, depression) and among healthy but stressed individuals.

Jon Kabat-Zinn, PhD, Center for Mindfulness at the University of Massachusetts Medical School

CONNECTION

Klein, Lauren. "Scratch a Happy Adult, Find a Socially Connected Childhood." *Greater*

Good Magazine, December 2013.
Lueras-Tramma, Nadine. "Feeling Connected Makes Us Kind." *Greater Good Magazine*, September 2011.
Flythe, Michelle. "Survival of the Social." *Greater Good Magazine*, September 2005.
Suttie, Jill. "How Social Connections Keep Seniors Healthy." *Greater Good Magazine*, March 2014.
George, Linda. "The Costs of Loneliness." *Greater Good Magazine*, December 2007.
Smith, Jeremy Adam. "How Love Grows in Your Body." *Greater Good Magazine*, February 2013.
Breines, Juliana. "Are Some Social Ties Better Than Others?" *Greater Good Magazine*, March 2014.

Remember Connection

Pavey, Louisa, Tobias Greitemeyer, and Paul Sparks. "Highlighting Relatedness Promotes Prosocial Motives and Behavior." *Personality and Social Psychology Bulletin* 37, no. 7 (April 2011): 905-917.

Some study participants reflected on a time when they felt a strong bond with someone else; other participants wrote about a time when they felt especially competent or autonomous. Compared with those in the other groups, the participants who reflected on their experience of closeness reported greater feelings of connectedness and concern for others. What's more, they also reported a stronger intention to carry out a variety of altruistic behaviors over the next six weeks, including giving money to charity and going out of their way to help a stranger in need. When they analyzed the data more closely, the researchers found that a greater desire to be kind depended on whether participants experienced greater feelings of connectedness to others after doing the writing exercise.

Compassionate Mediation

Weng, Helen, Andrew Fox, Alexander Shackman, Diane Stodola, Jessica Caldwell, Matthew Olson, Gregory Rogers, and Richard Davidson. "Compassion Training Alters Altruism and Neural Responses to Suffering." *Psychological Science* 24, no. 7 (July 2013): 1171-1180.

Study participants received either this compassion meditation training or a training aimed at mitigating negative emotion by helping people think differently about a negative event. Participants who completed two weeks of the compassion training demonstrated more altruism—they gave more money to a victim of unfair treatment. This altruistic behavior is a strong marker of compassion. What's more, the people who received the compassion training showed different brain activity in response to pictures of suffering: Their brains showed greater activity in regions known to be involved in understanding the suffering of others, regulating emotions, and experiencing positive feelings in response to a reward or goal. In this case, suggest the researchers, that goal was alleviating the suffering of someone in need.

Helen Weng, PhD, University of California, San Francisco

Center for Investigating Healthy Minds, University of Wisconsin, Madison

CURIOSITY

Cozolino, Louis. "Nine Things Educators Need to Know About the Brain." *Greater Good Magazine*, March 2013.
Harackiewicz, Judith, Kenneth Barron, John Tauer, and Andrew Elliot. "Predicting Success in College: A Longitudinal Study of Achievement Goals and Ability Measures as Predictors of Interest and Performance From freshman Year Through Graduation." *Journal of Educational Psychology* 94, no. 3 (September 2002): 562-575.
Wiswell, Albert and Thomas Reio Jr. "Field Investigation of the Relationship Among Adult Curiosity, Workplace Learning, and Job Performance." *Human Resource Development Quarterly* 11, no. 1 (January 2001): 5-30.

Noticing New Things

Langer, Ellen. *On Becoming an Artist: Reinventing Yourself through Mindful Creativity*. New

York: Ballantine Books, 2006.

In several different experiments—involving tasks as diverse as crocheting, watching football, eating different types of chocolate, and listening to non-preferred types of music—participants who were asked to notice new things about the experience reported liking it more afterward (and sometimes even said they were more likely to do it again) compared with others who weren't asked to notice new things. In addition to the research by Langer and colleagues cited in her book *On Becoming an Artist*, this exercise draws on suggestions from researcher Todd Kashdan, author of *Curious? Discover the Missing Ingredient to a Fulfilling Life*.

FORGIVENESS

Fincham, Frank, Steven Beach, and Joanne Davila. "Forgiveness and Conflict Resolution in Marriage." *Journal of Family Psychology* 18, no. 1 (April 2004): 72-81.

Howe, Donna. "Forgive Me?" *Greater Good Magazine*, June 2008.

Staub, Ervin, Laurie Anne Pearlman, Alexandra Gubin, and Athanase Hagengimana. "Healing, Reconciliation, Forgiving, and the Prevention of Violence After Genocide or Mass Killing: An Intervention and Its Experimental Evaluation in Rwanda." *Journal of Social and Clinical Psychology* 24, no. 3 (May 2005): 297-334.

Steps to Forgiveness

Baskin, Thomas and Robert Enright. "Intervention Studies on Forgiveness: A Meta-Analysis." *Journal of Counseling and Development* 82, no. 1 (Winter 2004): 79-90.

Researchers compared several studies that used Dr. Enright's "process model of forgiveness," similar to the steps outlined above. All the studies were done in a clinical setting including individual and group therapy. Therapies that used these methods were shown to be effective in increasing forgiveness, and in decreasing negative psychological states such as anxiety and anger. These were often long-term therapies, ranging from 6 to 60 weekly sessions, aimed at helping individuals cope with serious offenses.

Robert Enright, PhD, University of Wisconsin, Madison

WONDER

Wonder Narrative

Rudd, Melanie, Kathleen Vohs, and Jennifer Aaker. "Awe Expands People's Perception of Time, Alters Decision Making, and Enhances Well-Being." *Psychological Science* 23, no. 10 (August 2012): 1130-1136.

In three experiments, participants who were induced to feel awe, compared with other emotions, felt that they had more time available, were less impatient, were more willing to volunteer their time to help others, preferred experiences over material products, and reported greater life satisfaction.

Melanie Rudd, PhD, University of Houston

Wonder Story

Rudd, Melanie, Kathleen Vohs, and Jennifer Aaker. "Awe Expands People's Perception of Time, Alters Decision Making, and Enhances Well-Being." *Psychological Science* 23, no. 10 (August 2012): 1130-1136.

In three experiments, participants were induced to feel awe—such as by reading the story in this practice—as well as other emotions. People who experienced awe felt that they had more time available to themselves, were less impatient, were more willing to volunteer their time to help others, preferred having positive experiences over material products, and reported greater life satisfaction.

Melanie Rudd, PhD, University of Houston

An imprint of Mandala Earth
P.O. Box 3088
San Rafael, CA 94912
www.MandalaEarth.com

Find us on Facebook: www.facebook.com/MandalaEarth
Follow us on Twitter: @MandalaEarth

CEO: Raoul Goff
VP Publisher: Roger Shaw
Editorial Director: Katie Killebrew
Editor: Claire Yee
Editorial Assistant: Amanda Nelson
VP Creative: Chrissy Kwasnik
Art Director: Ashley Quackenbush
Designer: Amy DeGrote
VP Manufacturing: Alix Nicholaeff
Production Associate: Tiffani Patterson
Sr Production Manager, Subsidiary Rights: Lina s Palma-Temena

Mandala Earth would also like to thank Allie Kiekhofer.

ISBN: 978-1-64722-872-9

Mandala Earth, in association with Roots of Peace, will plant two trees for each tree used in the manufacturing of this book. Roots of Peace is an internationally renowned humanitarian organization dedicated to eradicating land mines worldwide and converting war-torn lands into productive farms and wildlife habitats. Roots of Peace will plant two million fruit and nut trees in Afghanistan and provide farmers there with the skills and support necessary for sustainable land use.

Manufactured in China

First printed in 2022.

10 9 8 7 6 5 4 3 2 1

2022 2023 2024 2025

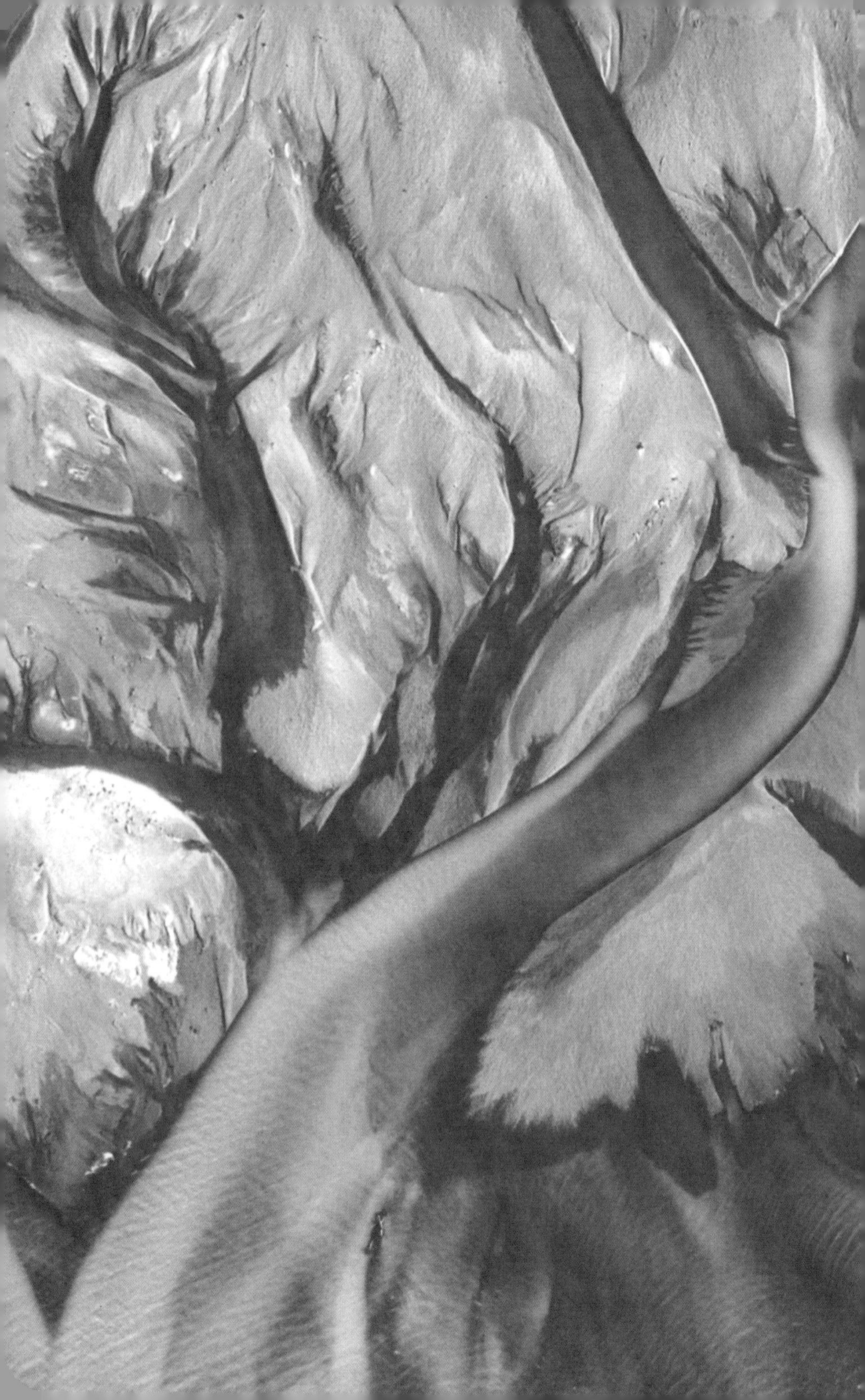